Ronald L. Greene

Classical Mechanics
with Maple

With 56 Figures

Springer

Ronald L. Greene, Ph.D
Department of Physics
University of New Orleans
New Orleans, LA 70148
USA

Library of Congress Cataloging-in-Publication Data
Greene, Ronald L.
  Classical mechanics with Maple / Ronald L. Greene.
  p.   cm.
  Included bibliographical references and index.
  ISBN 0-387-94512-1 (alk. paper)
  1. Mechanics—Data processing. 2. Maple (Computer file)
I. Title.
QC125.2.G72   1995
531'.1'078—dc20
                                    95-8322

Printed on acid-free paper.

Production managed by Natalie Johnson; manufacturing supervised by Jacqui Ashri.
Photocomposed using the author's TeX files.
Printed and bound by R.R. Donnelley and Sons, Harrisonburg, VA.
Printed in the United States of America.

9 8 7 6 5 4 3 2 1

ISBN 0-387-94512-1 Springer-Verlag New York Berlin Heidelberg

Preface

A few years ago I started playing around with the Maple V symbolic algebra software system. I was impressed with its ability to tackle significant physics problems with a rather small subset of the entire system. When I received my copy of Release 2 with its vastly improved and comfortably familiar output, I knew that this could be a significant contribution to the teaching of physics.

Like many physics professors, I had been thinking about how best to incorporate the use of computers into my courses. Traditional compilers, and even integrated programming environments such as Turbo Pascal, did not fit well into my courses because students spent so much time creating a working program that, at best, I could assign only a few problems to solve with the computer. In addition, I discovered that they learned more about programming than they did about physics. Some professors were recommending the use of spreadsheets in physics, but I felt that they were not powerful enough for upper division courses, and their use still required a significant amount of programming, although of a different sort.

Maple provides the possibility of a different approach. Not only does it have built-in numerical and graphical abilities, it also has symbolic capabilities which allow it to mesh well with our courses and their strong dependence upon symbolic mathematics. Just as important from the point of view of teaching physics is that Maple can be used in what computer scientists term a "non-declarative" programming mode; that is, we can tell the system *what* to do without having to become caught up in supplying the details of *how* to do it. This insures students can spend more time investigating physics and less on programming.

In my view, the way to incorporate computers into the undergraduate physics curriculum is to teach them to use a symbolic/numerical/graphical package such as Maple early in their studies, and then to use it as a unifying tool across all of their subsequent course work. This text attempts to illustrate how this can be done in the traditional junior-level classical mechanics course. It was written to serve primarily as a supplementary text to one of the standard classical mechanics texts. However, supported by additional theory and mathematics, it can be used as the primary text for such a course.

Contents

Introduction to Maple V

1.1 Basics

Maple V is an extensive software system capable of computing and manipulating data symbolically, numerically, and graphically. It is often generically referred to as a Computer Algebra System (CAS). Although it is a very powerful tool for mathematical manipulations, the sheer size of the program can be intimidating to someone considering using it in science or engineering. However, as will be seen in this text, even a small subset of Maple can be quite useful to the study of physics. Maple's symbolic abilities can reduce the amount of tedious algebra that is often necessary in solving physics problems. Furthermore, its numeric and graphical abilities allow us to go further into a given problem and extract physical understanding that is otherwise difficult to obtain. This book considers Maple's application specifically to the study of classical mechanics, but much of what is discussed is also applicable to other disciplines.

This first chapter presents a brief introduction to those aspects of Maple V we will use in studying classical mechanics. It is important to work through this chapter, trying out the examples, prior to moving into the rest of the text. It provides the Maple background necessary to understand the physics examples presented throughout the text. It is through working examples and problems that we gain expertise in using Maple as a problem solving tool.

This text does not detail the use of Maple on any particular computer system, although the descriptions generally apply to a version of Maple running under a graphical user interface, such as MS-Windows or Motif.

1.1.1 *Entering Expressions*

Maple is primarily an interactive tool in which we type and enter commands, causing the computer to respond to those commands. The program displays a prompt (usually a > sign) when it is waiting for input. The commands that we type are typically algebraic expressions, assignments, or function calls. Maple is case-sensitive; we must pay attention to this particularly when issuing Maple commands. Statements should be terminated by a semicolon (;) or a colon (:) to let the Maple parser know that it has reached the end of the command. A semicolon causes Maple to print out the result of the command, whereas a colon suppresses the output. We can extend a command over several lines by hitting the <Enter> key to start a new line. Maple attempts to execute the command only when the ';' or ':' is found. We can also put multiple commands on one line, as long as each command ends with a valid terminator.

The following are simple examples of inputs and Maple responses. The > prompt is not part of the input.

```
> 3*4;
```
$$12$$

```
> 1 + 2 + 3 + 4 + 5
>   + 6 + 7 + 8;
```
$$36$$

```
> 2*x - 5 + x;
```
$$3x - 5$$

```
> -2^4;     (-2)^4;
```
$$-16$$
$$16$$

```
> 4!;
```
$$24$$

Maple does simple arithmetic both numerically, as in $3*4$ returning 12, and symbolically, as in combining $2x$ and x to get $3x$. In addition to the four primary arithmetic operations +, -, *, and /, Maple can exponentiate (^) and take factorials (!). Other operations can be performed with function calls.

If we make a mistake when typing an expression, we can backspace to the point of error and retype from there, or use the arrow keys or mouse to move the cursor to the appropriate place to delete characters or insert additional ones. The cursor does not need to be at the end of the line in order to enter a command. As long as the expression (including those formed from multiple-line expressions) is properly terminated, Maple processes it. If there is a syntax error in the statement as we have entered it, Maple returns a comment to that effect, with an indication as to where the error occurred. We can then move the cursor back up and make the appropriate correction.

If our command is one that Maple does not recognize (*e.g.*, a misspelling of a command name or a capital letter that should have been lower case, or a function not yet defined), Maple usually returns the input as is, with no explanation. This may be puzzling the first few times it happens, so keep it in mind.

There is a well-defined order of precedence for the Maple arithmetic operators. Generally, the precedence is the same as that of most common programming languages, such as Pascal, Fortran, or C. As seen in the fourth example above, parentheses can be used to arrange the desired order of operations. The precedence rules, as well as other information about Maple operators can be obtained by entering the command ?precedence at the Maple prompt. Help on any command and many general topics can be accessed on-line by preceding the topic with a question mark. Help requests do *not* have to be terminated by a semicolon or colon. Help may also be obtained by clicking on "Help" in the Maple interface window.

There are often times when we need to refer to the immediately previous result. Maple allows us to refer to the result of the last *evaluated* expression with a double quotation mark ("). Expressions resulting in syntax errors do not count. Similarly, two double quotes and three double quotes refer to the second to last and third to last evaluated expressions, respectively.

```
> (x+3)^2;
```
$$(x + 3)^2$$

```
> " - 5;
```
$$(x + 3)^2 - 5$$

```
> "";
```
$$(x + 3)^2$$

Note that " refers to the immediately previous result regardless of whether it was printed out or not. So, for example, had the first input been terminated with a colon rather than a semicolon, the results of the second and third inputs and outputs would not have changed.

1.1.2   Fundamental Data Types

Maple deals with several kinds of numbers. Integers, rational numbers, irrational numbers, and floating-point numbers are fundamental. In addition, the program can manipulate complex numbers, whose real and imaginary parts can be any of the fundamental types. Except for floating-point numbers (*i.e.*, numbers containing a decimal point), Maple arithmetic is exact to however many digits are necessary, assuming the computer has sufficient memory and we have sufficient time. On those occasions that call for a floating-point approximation to an exact result, we can call the evalf command with the desired expression as its argument. For example,

> 352/1200;

$$22/75$$

> evalf(");

$$.2933333333$$

The accuracy of floating-point arithmetic is controlled by the global variable Digits, and can be increased or decreased by changing its value.

If there are floating-point numbers in an expression that we enter, Maple automatically does some type changing and floating-point evaluation of the expression. For example,

> 1.1 + 22/75;

$$1.393333333$$

but

> 1.1 + 2^(1/2);

$$1.1 + \sqrt{2}$$

In this latter example, we must apply the evalf command if we want a floating-point result.

For complex numbers, Maple uses the symbol I to represent $\sqrt{-1}$. Hence,

> (3 + 5*I/12)*(-2 + I);

$$-\frac{77}{12} + \frac{13}{6}I$$

In addition to numbers, Maple can manipulate symbols, such as x in several of the examples above. Maple's ability to do algebra and calculus symbolically is one of the components that makes the program such a powerful tool for doing physics.

1.1.3 Basic Mathematical Functions

Maple contains knowledge of many mathematical functions, including how to evaluate them for specific argument values, and how to manipulate them symbolically — to differentiate and integrate, apply identities, *etc.* Some of Maple's functions that are commonly used in physics are listed in Table 1.1. In addition, the inverse trigonometric and hyperbolic functions are available with the corresponding names prepended by the characters 'arc': arcsin, arcsech, *etc.*

Any of these functions may be called by placing its argument in parentheses; for example,

> exp(-alpha*x);

$$e^{(-\alpha x)}$$

Description	Maple Name
Absolute value	abs
Square root	sqrt
Exponential function	exp
Natural logarithm	ln, log
Base 10 logarithm	log10
Trigonometric functions	sin, cos, tan, csc, sec, cot
Hyperbolic functions	sinh, cosh, tanh, csch, sech, coth
Bessel functions	BesselI, BesselJ, BesselK, BesselY
Error functions	erf, erfc
Gamma function	GAMMA

TABLE 1.1. Common Mathematical Functions.

```
> ln(exp(2)) + sin(Pi/2);
```
$$3$$

```
> arctan(1);
```
$$\tfrac{1}{4}\pi$$

```
> erf(2);    erf(2.0);
```
$$\mathit{erf}\,(2)$$
$$.9953222650$$

We note that in the first example Maple recognizes the symbol alpha to be the Greek letter α and displays it as such in the output. The capitalization of the English spelling of a Greek letter determines the case of the output, with one exception. The letter π is the exception. Maple recognizes Pi as the mathematical constant 3.14159.... It interprets pi as a symbol whose output is the Greek letter π, which is *not* the same as the mathematical constant. This can be highly confusing; it is probably best to avoid using pi completely, and pay special attention that Pi is always capitalized.

Another potentially confusing Greek letter is γ. Maple uses gamma (in input) and γ (in output) for the Euler constant. In this text we avoid using it for an arbitrary symbolic variable, as is commonly done in other classical mechanics texts, because Maple will not warn us about it unless we try to make an assignment to the symbol.

In addition, there are two capital Roman letters that we should be careful about using. E and I are both Maple constants that should not be used as symbolic variables. E is the base of natural logarithms, and I is $\sqrt{-1}$. These symbols are commonly used in physics to represent energy and moment of inertia, but we should avoid using them. Particularly vexing is the fact that Maple provides no warning about their use, unless we try to assign values to them, at which point we may find it necessary to re-enter all our previous inputs after replacing E or I with different symbols.

Returning to the examples above, we note in the second example that the argument of a function can be another function call, and that Maple may make some automatic simplifications to return a simple and exact answer. Also, from the fourth example, we see that if Maple is given a non-floating-point argument of a function it cannot evaluate, it simply returns the input, whereas if given a floating-point value it evaluates the function, returning a floating-point value.

There are many other mathematical functions in Maple, as well as commands that allow us to manipulate expressions. Some from the latter category are presented later in this chapter. For a complete listing of the commands initially known to Maple we can type ?inifcns at the prompt. For other commands that can be loaded, refer to *The Maple Handbook*[14].

1.1.4 *Variables*

Maple is capable of algebraically manipulating symbols, which represent variables. In previous examples the variable x was "unassigned," in that it had no value other than its name. Such unassigned variables are commonly used in Maple as algebraic unknowns, or for such things as "dummy" indices or variables in sums or integrals, or formal parameters in function definitions. On the other hand, it may be convenient to assign the result of an operation to a variable so that it can be referred to in future calculations. This is done with the Maple assignment operator (:=). Variables on the left side of the := are assigned the value of whatever the right side evaluates to. Wherever that variable appears on the right side of a subsequent expression, it is replaced by its value (with a few exceptions, discussed later). For example,

```
> x := 3*5 - 1/2:
> x^2;
```
$$\frac{841}{4}$$

```
> y := 1 + u^2;
```
$$y := 1 + u^2$$

```
> u := 3:   y;
```
$$10$$

Notice particularly in the last two inputs that when y was initially assigned, u was unassigned, so that y's value was given in terms of the symbol u. After assigning a value to u, when we ask for y's value the value of u is substituted to give the final result. This is an example of "full evaluation," which normally is performed to the expression on the right side of the :=. Under full evaluation all variables in the expression are replaced by their values, and if the result contains any assigned variables they are replaced by their values, and so on until there are no more assigned variables remaining.

An important thing to notice is that internally the value of y is still $1+u^2$. We can see this by unassigning the variable u and asking for the value of y again. A variable can be unassigned or cleared by assigning it its own name:

> u := 'u';

$$u := u$$

The single quotes tell Maple not to evaluate what is inside; this is one of the exceptions to the full evaluation rule. If we ask for the value of y as before, we get

> y;

$$1 + u^2$$

We will find that this behavior of changing the value of a stored variable under full evaluation, but not changing its internally stored value, is very convenient for studying the behavior of a given physical system under various values of its parameters.

A word of caution is in order, however. If we are not careful, full evaluation can lead to a situation of infinite recursion, which will force an abort of the Maple session. For example, suppose t is an unassigned variable. If we make the assignment

> t := t + 1:

Warning recursive definition of name

and then try to use t, Maple will be caught up substituting t+1 for t, and then substituting t+1 for the t in that result, and so on until it overflows its allotted memory. In a simple case like this, Maple recognizes the potential recursion and issues a warning when we make the assignment. However, it is possible, without getting any warning, to make a circuitous series of assignments that trigger an infinite recursion when used.

When moving from one problem to another, it is best to clear previously assigned variables. We have seen one way to do this — assigning its own name to each variable by enclosing the name in single quotes — but this can be tedious with a lot of variables. The easiest way to clear all variable assignments is to invoke the Maple restart command. Typing

> restart;

begins a completely new session. We must be sure that is what we want to do since all our work to that point is no longer accessible. The remaining examples in the first two chapters of this text explicitly unassign variables or call restart at the beginning. However, unless otherwise stated, each example in subsequent chapters assumes no remnants from previous Maple sessions.

1.1.5   Sequences, Lists, and Sets

Expressions separated by commas make up a *sequence*. In general, the arguments of Maple commands are sequences, although in many cases the sequence may have only one element. In addition, some Maple commands return sequences. For example, as seen in the next section, in solving an equation that has multiple roots, Maple returns a sequence of roots. We may then select a particular root to work with. There are special commands for creating sequences, converting a mathematical expression into a sequence of its parts, and other manipulations, but except for the extraction of a particular element, they are not used in this text. Explanations and examples of other sequence operations may be found in general Maple texts.

A *list* is a sequence enclosed by brackets, as in

```
> restart;
> list1 := [1, 2, 3];
```
$$list1 := [1, 2, 3]$$

```
> list2 := [1+x², 1-x²];
```
$$list2 := [1 + x^2, 1 - x^2]$$

Similarly, a *set* is a sequence enclosed by braces, such as

```
> set1 := {1, 2, 3};
```
$$set1 := \{1, 2, 3\}$$

```
> set2 := {1+x², 1-x²};
```
$$set2 := \{1 + x^2, 1 - x^2\}$$

There are two major differences between the two structures. The first is that a list is strictly ordered, whereas the order of elements within a set is arbitrary. This distinction shows up in a number of places in Maple. For example, as we will see in the next section, when solving a set of linear algebraic equations Maple returns a set of solutions for the unknowns. The ordering of the unknowns in the solution set is internally determined by Maple, and can vary from one instance to the next, even when solving the same equations. The other major difference between a list and a set is that a set does not contain any duplicated elements, whereas two or more elements of a list may be identical. Maple automatically eliminates duplicate elements in a set.

Individual elements of a list or set may be obtained with the selection operator, [ ], as with

```
> list2[1];    set2[2];
```
$$1 + x^2$$
$$1 - x^2$$

1.2   Algebraic Equations

One of the most useful features of Maple is its ability to solve simultaneous algebraic equations, both linear and, within limits, nonlinear. Such equations occur frequently in physics; examples are given throughout this text, especially in Chapter 2. It is relatively easy to deal manually with two such equations, or perhaps even three. However, when we get up to four or more the tedium of solving the set and the likelihood of making algebra mistakes discourages attempting the solution. Maple solves such sets of equations with ease and accuracy through the solve command. The first argument of solve is the algebraic equation or set of equations to be solved, and the second argument is the variable or set of variables to be solved for.

Let us consider two single-equation examples. The first is a general linear equation in the variable x:

```
> restart;
> eq := (1-a^2)*x + b = c*y;
```

$$(1 - a^2)x + b = cy$$

```
> solve(eq, x);
```

$$-\frac{b - cy}{1 - a^2}$$

It is very important to understand the distinction between the = and := signs in the second line above. The := is the assignment operator. Everything to its right is an expression, which is assigned to the variable eq for convenience in further manipulation. In this case the assigned expression happens to be an equation, which is indicated by the = sign. For this example we can avoid introducing the variable eq by entering the equation directly as the first argument of the solve command, or by entering the equation on the second line (without the assignment) and using " for the first argument of solve. However, we will see in future examples that assigning equations to variables is often useful, from the point of view of convenience, as well as clarity.

We note that Maple solves equation eq for x without concern for the possibility that $1 - a^2$ might be zero. It is our responsibility to consider such possibilities. Also note that solve does not assign a value to the variable x. If desired, we can make the assignment with the statement

```
> x := ":
```

The second example treats a quadratic equation. It is an energy conservation equation for an object dropped from a height h near the earth's surface.

```
> restart;
```

> 1/2*m*v^2 = m*g*h;

$$\tfrac{1}{2}mv^2 = mgh$$

> solve(", v);

$$\sqrt{2}\sqrt{gh}, \ -\sqrt{2}\sqrt{gh}$$

This time the result of the solve command is a two-element sequence, representing two mathematically acceptable solutions to the equation. The physically acceptable solution is the positive one since v represents the speed of the object just before hitting the ground. We choose the first solution by using the selector [1] as in

> v := "[1];

$$v := \sqrt{2}\sqrt{gh}$$

It is only after this assignment that v has the value shown; *the solve command does not assign any values to the variables it solves for*. Also remember that the order of the solutions can vary. We need to pay attention to the order of our result to insure that we select the solution that we want.

solve can get closed form solutions to the roots of any polynomial through degree four. A well-known mathematical theorem states that it is not possible to find the roots of a general polynomial of degree higher than four. Consequently, if we ask Maple to find the solutions to a general higher-order polynomial it returns a result in terms of a RootOf expression, which is a representation for the unknown solutions. This usually is not very helpful, although Maple can do certain manipulations with RootOf expressions. RootOf expressions are discussed to some extent in Chapter 2 in the context of another conservation of energy problem.

The solve command can also obtain solutions to some non-polynomial, non-linear equations as illustrated by

> x := 'x':
> solve(sqrt(1+x^2)=2*x, x);

$$\frac{1}{6}\sqrt{4}\sqrt{3}$$

but of course cannot solve transcendental equations.

As a final example of using solve in this introduction (see other examples throughout the text), consider three linear equations in the variables x, y, and z.

> restart;
> eq1 := a*x + 2*y - 5*z = 0:
> eq2 := x - 1/2*y + z = 2:

> eq3 := 2*x + 3*y - z = 0:

We solve these three equations simultaneously and assign the result to the variable sol with the following command:

> sol := solve({eq1,eq2,eq3}, {x,y,z});

$$sol := \left\{ x = 52\frac{1}{5a+28},\ y = 4\frac{-10+a}{5a+28},\ z = 4\frac{-4+3a}{5a+28} \right\}$$

Maple is a large program and, like most (perhaps all) large programs, it has some bugs. Consequently, it is always prudent to check our solutions. The most convenient way to do so is to substitute them back into the original equations, simplify, and verify that the results are identities. Maple's subs command will perform the substitution. In checking, the arguments to subs are the solution set and the equation set:

> subs(sol, {eq1,eq2,eq3});

$$\left\{ 52\frac{a}{5a+28} + 8\frac{-10+a}{5a+28} - 20\frac{-4+3a}{5a+28} = 0, \right.$$

$$104\frac{1}{5a+28} + 12\frac{-10+a}{5a+28} - 4\frac{-4+3a}{5a+28} = 0,$$

$$\left. 52\frac{1}{5a+28} - 2\frac{-10+a}{5a+28} + 4\frac{-4+3a}{5a+28} = 2 \right\}$$

This complicated mess is made intelligible by

> simplify(");

$$\{2 = 2,\ 0 = 0\}$$

The identities indicate that the solutions are correct. Because the result is a set, two $0 = 0$ elements have been merged into one.

It is usually convenient for checking to combine the subs and simplify commands into a single expression,

> simplify(subs(sol,{eq1,eq2,eq3}));

$$\{2 = 2,\ 0 = 0\}$$

This Maple idiom is very commonly used to check solutions, and is well worth remembering.

Should you have confidence in Maple-generated solutions being checked with Maple? Probably, since a different part of the program code is used in the checking phase than in the solution phase. It is very unlikely that errors in each part end up canceling each other's effect. Just as for humans, it is usually easier (and thus less error-prone) for Maple to check a solution by substitution than to generate a solution in the first place.

As in the previous examples, Maple made no assignments to the variables x, y, and z when it found the solution to the three equations. The solution sol is a set of equations, as indicated by the = signs, rather than assignments, which would use the := operator. We can assign each of the variables on the left sides of the equations in sol their corresponding values from the right side with the assign command. It takes one argument, either an equation with a variable name on the left side of the = sign and an expression on the right side, or a set of such equations. The latter form is what was returned from the solve command and assigned to sol, so the statement

> assign(sol);

makes the appropriate assignments to x, y, and z.

> x;   y;   z;

$$52\frac{1}{5a+28}$$

$$4\frac{-10+a}{5a+28}$$

$$4\frac{-4+3a}{5a+28}$$

If the only non-numeric quantities in the equations to be solved are the unknowns, the Maple fsolve command can obtain real floating-point solutions. This is useful, for example, in finding the zero crossings of given functions, as in

> x := 'x':
> x^5 - 3*x^3 + 4*x^2 - 2;

$$x^5 - 3x^3 + 4x^2 - 2$$

> fsolve("=0, x);

$$-2.157508716, \ -.5986128337, \ 1.$$

The command fsolve can produce numerical solutions for problems that solve cannot handle, such as the roots of polynomials of degree higher than four, or transcendental equations.

1.3   Calculus and Differential Equations

Maple has a great deal of knowledge about the mechanics of calculus. It can symbolically differentiate, integrate, find limits, and obtain series expansions of expressions. In addition, it can symbolically solve many kinds of linear differential equations and some non-linear ones. Maple can also numerically solve integrals and differential equations. All of these abilities are quite useful in studying physics.

1.3.1   Differentiation and Integration

Maple's command for symbolic differentiation is diff. The first argument to diff should evaluate to the expression to be differentiated. The remaining arguments are the variables with respect to which differentiation is performed. So, for example,

```
> restart;
> expr := x^4/y;
```

$$expr := \frac{x^4}{y}$$

```
> diff(expr, x);
```

$$4\frac{x^3}{y}$$

```
> diff(expr, x, x);
```

$$12\frac{x^2}{y}$$

```
> diff(expr, x, y, y, y);
```

$$-24\frac{x^3}{y^4}$$

(There is an alternate way of expressing higher-order derivatives which uses the sequence operator $. See references if interested.)

Indefinite integration is handled in a similar manner with the int command, except that multiple integrations require multiple calls to int. With the same value for expr as above, we find

```
> int(expr, x);
```

$$\frac{1}{5}\frac{x^5}{y}$$

```
> int(int(expr,x), y);
```

$$\frac{1}{5}x^5 \ln(y)$$

In the latter instance, the integration over x is performed first since in nearly all cases Maple evaluates the arguments to a command prior to invoking it.

For definite integration the second argument to int is expanded to a form which includes the upper and lower limits on the integration, as in

> int(expr, y=1..5);

$$x^4 \ln(5)$$

> int(sin(a*x)^2, x=0..2*Pi/a);

$$\frac{\pi}{a}$$

> int(exp(-t^2), t=-infinity..infinity);

$$\sqrt{\pi}$$

If we try a very similar problem,

> int(exp(-a*t^2), t=-infinity..infinity);

$$\int_{-\infty}^{\infty} e^{(-at^2)} \, dt$$

we find that Maple is unable to perform the integration. This result may be surprising in light of the success with the same integral with $a = 1$; however, it illustrates a very important point. The result of the integral depends on the properties of a. In physics we usually implicitly assume that our parameters are real, and often positive. Maple does not assume this to be the case unless we tell it to. We do so with the assume command. Thus, we get the expected answer as follows:

> assume(a > 0):
> int(exp(-a*t^2), t=-infinity..infinity);

$$\frac{\sqrt{\pi}}{\sqrt{a^{\tilde{}}}}$$

The ~ symbol is used by Maple to remind the user that the variable a has an assumed property. Additional properties can be added with the additionally command. See ?assume for details of assumable properties and related commands. Assumptions on a given variable can be removed by unassigning the variable.

There are several options that can be added as a third argument to the int command. Entering ?int gives information about some of these; see *The Maple Handbook*[14] for further information.

The Maple integrator is quite powerful. It can express many integrals that cannot be evaluated in closed form in terms of special functions. For example,

> int(exp(-t^2), t=0..x);

$$\frac{1}{2}\sqrt{\pi}\,\mathrm{erf}(x)$$

> int(1/sqrt((x^2-1)*(x^2-2)), x=0..1);

$$\frac{1}{2}\sqrt{2}\,LegendreF(1,\tfrac{1}{2}\sqrt{2})$$

The function in the latter example is an elliptical integral of the first kind. If Maple cannot evaluate the integral it simply returns the entered expression, printing it out in standard mathematical form.

We can request numerical integration by supplying the int expression as an argument to the evalf command. Of course, all parameters in the integrand must have numeric values. As an example,

> int(exp(-x^3), x=0..1);

$$\int_{0}^{1} e^{(-x^3)}\,dx$$

> evalf(");

$$.8075111821$$

This integral cannot be performed analytically, so Maple returns it unevaluated. In this case the call to evalf invokes the numerical integrator. If we know that Maple cannot symbolically evaluate an integral, we can prevent its attempt (and thus save time) by using Int in place of int. Int is the "inert" form for the integration command. See ?int[numeric] and ?value for further information.

1.3.2   Solving Differential Equations

The first step in solving one or more differential equations encountered in a given problem is to enter it, or them, into Maple. As with algebraic equations, it is often convenient to give each equation a label. Once this is done the equation(s) can be solved, presuming Maple can solve it (them), with the dsolve command. Assigning the result of dsolve to a variable makes it easy to check the solution by substituting back into the equation. Study the following example.

> restart;
> eq := diff(x(t), t) + alpha*x(t) = 0;

$$eq := \left(\frac{\partial}{\partial t}x(t)\right) + \alpha\,x(t) = 0$$

> sol := dsolve({eq, x(0)=x0}, x(t));

$$sol := x(t) = e^{(-\alpha t)}\,x0$$

> simplify(subs(sol,eq));

$$0 = 0$$

In the second statement, derivatives in the equation are entered using the diff command, and the dependent variable is explicitly shown to depend on the independent variable, as in x(t). In the third input, the dsolve command is called. Its first argument is the set of equations to solve — in this case the differential equation and the single initial condition on x(t). Had we not specified the initial condition, the solution would have contained an arbitrary constant, such as _C1. An equation is returned from the dsolve call, suitable for using in subs to substitute back into the original differential equation, as done in the fourth input line. This checking procedure is the same as used earlier for checking solutions to algebraic equations.

Solving the equation does not assign a value to x(t). That can be done with the assign command, just as for algebraic equations.

> assign(sol);

Also note that x(t) is a Maple variable, not a function, despite its appearance. In particular, we cannot obtain the value of x at $t = 0$ simply by a reference to x(0). The x(0) in the initial condition equation of the dsolve call is misleading in that regard. The creation of functions is discussed a little later in this chapter.

It is usually better not to assign values to variables from a solution unless you anticipate doing additional manipulations on them, such as checking limiting values. One reason for this is that, under full evaluation, assigning a value to x(t) causes other changes where x(t) appears. For example, look at the effect of the above assignment on the equation eq by typing

> eq;

The equation can be restored to its original form by unassigning x(t),

> x(t) := 'x(t)';

Second- and higher-order differential equations may be solved in a similar way. Of course, more initial conditions must be specified if the solution is not to contain arbitrary constants. Observe the way that an initial condition on the first derivative is specified in the following example. The example is the equation of motion for a harmonic oscillator with initial position x_0 and initial velocity v_0.

```
> restart;
> eq := diff(x(t), t, t) + omega^2*x(t) = 0;
```

$$eq := \left(\frac{\partial^2}{\partial t^2} x(t) \right) + \omega^2 x(t) = 0$$

```
> sol := dsolve({eq, x(0)=x0, D(x)(0)=v0}, x(t));
```

$$sol := x(t) = \frac{v0 \sin \omega t}{\omega} + x0 \cos \omega t$$

```
> simplify(subs(sol,eq));
```

$$0 = 0$$

Think of the D(x)(0) in the first argument of dsolve above simply as notation indicating the first derivative of x with respect to t evaluated at t = 0. The D operator has other uses which are not discussed in this text. We use it only as a way of specifying initial conditions on the first derivative of a function. Enter ?D further information.

The dsolve command can have a third, optional, argument, which may take any of several values. See ?dsolve for details. One of the options is numeric, which instructs Maple to try to solve the differential equation and its initial conditions numerically. All parameters, such as the ω in the example above, must be given numerical values for this to be successful. The result of a dsolve call with the numeric option is a function. This function can then be evaluated at the desired points. For example, consider the non-linear equation below, which has no closed-form solution.

```
> diff(x(t), t, t) + sin(x(t)) = 0;
```

$$\left(\frac{\partial^2}{\partial t^2} x(t) \right) + \sin(x(t)) = 0$$

```
> sol := dsolve({", x(0)=1, D(x)(0)=0}, x(t), numeric):
> sol(.5);
```

$$\left[t = .5, \ x(t) = .8960325471229542, \ \frac{\partial}{\partial t} x(t) = -.4108785137543359 \right]$$

The result of the solution function call is a list of equations showing floating-point values for the independent and dependent variables, and the derivative of the dependent variable. We may use subs to extract the x(t) part:

```
> subs(sol(.5), x(t));
```

$$.8960325471229542$$

This works because sol evaluated at a particular time, such as sol(.5), returns a list of equations in the form required by subs. Wherever x(t) appears in the second argument of subs it is replaced by the number on the right side of the equation x(t)=... in the list returned by sol(.5).

Alternatively, the value of x(t) may be obtained using the rhs (right-hand-side) command in conjunction with the selection operator [ ], as in

> rhs(sol(.5)[2]);

$$.8960325471229542$$

1.3.3   Limits and Series

Maple can calculate limits and series with the limit and series commands. These are especially useful for finding limiting cases or small parameter expansions of more complicated physics problems. The examples below illustrate their use:

> limit(sin(x)/x, x=0);

$$1$$

> series(sin(x)/x, x);

$$1 - \frac{1}{6}x^2 + \frac{1}{120}x^4 + O\left(x^5\right)$$

> series(exp(x), x=delta, 3);

$$e^\delta + e^\delta (x - \delta) + \frac{1}{2}e^\delta (x - \delta)^2 + O\left((x - \delta)^3\right)$$

As you can see, the point about which the expansion is taken defaults to 0, and the order of the expansion can be controlled by a third, optional parameter.

The output of the series command can be truncated to a polynomial so that it can be used in further calculations with the convert(..., polynom) command. See ?convert[polynom].

1.4   Simplification and Manipulation of Results

After working with Maple for even a short while we can see some very long and involved expressions. The simplify command does a reasonable job in making such expressions more manageable. In addition to the simple call we have used thus far, simplify can take one or more of several optional arguments. For example, simplify(...,symbolic) performs simplification of radicals ignoring the possibility that the variables inside the radicals may be negative. simplify(...,assume=real) and simplify(...,assume=positive) perform simplifications assuming that all variables in the expression are real or positive, respectively. (See ?simplify or *The Maple Handbook*[14] for further details.) Even so, sometimes we need to provide Maple with more explicit instructions to get it to do what we want. Other times we may wish to extract a part of the expression, such as the numerator of a rational expression or the right side of an equation. Maple has a number of commands designed to do specific tasks of this sort. Some of them are examined in this section. All of these commands, when given a variable which has been assigned an expression, will return a modified expression, but unless the variable is reassigned the result, its value is not changed. For example,

> a := (x^2-4)/(x+2);

$$a := \frac{x^2 - 4}{x + 2}$$

> simplify(a);

$$x - 2$$

> a;

$$\frac{x^2 - 4}{x + 2}$$

Note the value of a is not changed by simplify(a).

1.4.1   factor, expand, normal

The Maple command factor can be used to factor polynomials and expressions containing polynomials. Its first argument is the expression. A second, optional argument can be added to cause factorization over an algebraic extension field, such as a radical.

> factor(x^4-16);

$$(x - 2)(x + 2)(x^2 + 4)$$

> factor(x^3+3, 3^(1/3));

$$(x^2 - 3^{1/3}x + 3^{2/3})(x + 3^{1/3})$$

The command expand distributes multiplications and divisions over additions and subtractions. Thus, it can reverse the effect of factor for a polynomial. However, for general expressions, it also applies multiple-angle rules to trigonometric and hyperbolic functions, and sum-of-argument rules to exponentials. Thus,

> expand((x-2)*(x+2)*(x^2+4));

$$x^4 - 16$$

> expand(sin(alpha+beta));

$$\sin(\alpha)\cos(\beta) + \cos(\alpha)\sin(\beta)$$

> expand(exp(y + ln(x+1)));

$$e^y x + e^y$$

normal puts an expression into the form of numerator over denominator, and divides out the greatest common divisor of numerator and denominator. For example,

> 2 + (x^2 + 3*x + 2)/((x+2)*(x-1));

$$2 + \frac{x^2 + 3x + 2}{(x + 2)(x - 1)}$$

> normal(");

$$\frac{3x - 1}{x - 1}$$

simplify performs the task of normal in addition to other manipulations. As a result, simplify is used more often, even in cases where normal would be sufficient.

1.4.2 collect, sort, coeff

When applied to a polynomial (its first argument), collect returns a polynomial whose terms are collected by powers of the unassigned variable in the second argument position. For example,

> restart;
> p := expand((1+a-x)*(x+a)^2);

$$p := x^2 + 2xa + a^2 - ax^2 + xa^2 + a^3 - x^3$$

> collect(p, x);

$$-x^3 + (-a + 1)x^2 + (a^2 + 2a)x + a^3 + a^2$$

If the expression is a multivariate polynomial, the second argument of collect can be a list or set of unassigned variables. A third, optional argument can be added which specifies the form for the collection. (The third argument is effective only if the second argument is a list of variables.) Two possible forms are allowed: recursive (default) or distributed. The latter is useful for identifying terms with particular powers of each of two or more variables.

> collect(p, {x,a}, distributed);

$$x^2 + 2xa + a^2 - ax^2 + xa^2 + a^3 - x^3$$

Another optional argument is a simplification function to be applied to the coefficients of the various powers of the variable being collected. For example, if we want to factor the coefficients of x,

> collect(p, x, factor);

$$-x^3 + (-a + 1)x^2 + a(a + 2)x + a^2(a + 1)$$

The collect command can be used in many ways that are not discussed in this text. For example, its second argument can be a subexpression rather than a simple variable. See ?collect or *The Maple Handbook*[14] for further information.

The terms of a Maple polynomial are not ordinarily put into any particular order. To force ordering, we use the sort command. See ?sort. Two examples of using sort are given below. In the first, the expression is treated as a univariate polynomial in x, and terms are sorted in order of descending powers of x. In the second example the same expression is treated as a multivariate polynomial in x and y, and sorted in descending order in the total degree of the powers of x and y.

> p := expand((1+2*x)*(x+y)^2);

$$p := x^2 + 2xy + y^2 + 2x^3 + 4x^2y + 2xy^2$$

> sort(p, x);

$$2x^3 + x^2 + 4yx^2 + 2yx + 2y^2x + y^2$$

> sort(p, {x,y}, tdeg);

$$2x^3 + 4x^2y + 2xy^2 + 2xy^2 + x^2 + 2xy + y^2$$

Once we have a polynomial in an expanded or collected form, we can extract the coefficient of individual powers of the variables by using the coeff command. For example, to get the coefficient of the x^2 term in the polynomial p above, we use

> coeff(p, x, 2);

$$4y + 1$$

The result, of course, can be assigned to a variable if further processing is desired. An alternate form, which uses the fact that the second argument of coeff can be a subexpression and the third argument is by default 1, is

> coeff(p, x^2);

$$4y + 1$$

combine call	Example transformation
combine(expr, exp)	$e^x e^y \to e^{x+y}$
combine(expr, ln)	$\ln(x) + \ln(y) \to \ln(xy)$
combine(expr, power)	$x^a x^b \to x^{a+b}$ $\qquad (x^a)^b \to x^{ab}$
combine(expr, trig)	$\sin(x)\cos(y) \to \frac{1}{2}\sin(x+y) + \frac{1}{2}\sin(x-y)$
	$\sinh(x)\cosh(y) \to \frac{1}{2}\sinh(x+y) + \frac{1}{2}\sinh(x-y)$

TABLE 1.2. Effects of combine.

1.4.3 combine

In many ways combine acts oppositely to expand. The first argument of combine is the expression to be manipulated. The second argument, which is optional, describes a specific class of transformations that it will perform. Table 1.2 illustrates the combine calls that are most useful to us and examples of the transformation rules that are applied. So, for example,

> combine(cos(alpha)*cos(beta), trig);

$$\tfrac{1}{2}\cos(\alpha - \beta) + \tfrac{1}{2}\cos(\alpha + \beta)$$

It is well worth the time to experiment with the kinds of transformations described in Table 1.2 and to commit them to memory.

1.4.4 convert

Finally, we consider convert, a command that allows conversion between equivalent forms, such as trigonometric forms to exponentials or the reverse. Like combine, convert takes two arguments, the first being the expression to be manipulated and the second a descriptor controlling the type of transformation. Table 1.3 gives a summary of conversions available.

The following Maple session gives an example of using combine and convert together to simplify a trigonometric expression.

> expr := cos(alpha)^2*cos(beta)^2 - 2*cos(alpha)*cos(beta)*
>   sin(alpha)*sin(beta) + sin(alpha)^2*sin(beta)^2;

$$expr := \cos(\alpha)^2 \cos(\beta)^2 - 2\cos(\alpha)\cos(\beta)\sin(\alpha)\sin(\beta)$$
$$+ \sin(\alpha)^2 \sin(\beta)^2$$

> simplify(expr);

$$2\cos(\alpha)^2 \cos(\beta)^2 - 2\cos(\alpha)\cos(\beta)\sin(\alpha)\sin(\beta)$$
$$+ 1 - \cos(\beta)^2 - \cos(\alpha)^2$$

> combine(expr, trig);

$$\tfrac{1}{2}\cos(2\alpha + 2\beta) + \tfrac{1}{2}$$

convert call	Transformation
convert(expr, exp)	trigonometric forms to exponentials
convert(expr, ln)	inverse trigonometric forms to logarithms
convert(expr, expln)	combination of exp and ln conversions
convert(expr, trig)	exponential forms to sin, cos, sinh, cosh
convert(expr, sincos)	trigonometric forms to sin, cos, sinh, cosh
convert(expr, tan)	trigonometric forms to tangents
convert(expr, expsincos)	trigonometric forms to sin, cos;
	hyperbolic forms to exponentials
convert(expr, factorial)	Γ functions, binomial coefficients
	to factorials
convert(expr, GAMMA)	factorials, binomial coefficients
	to Γ functions

TABLE 1.3. Effects of convert.

> convert(", tan);

$$\frac{1}{2}\frac{1 - \tan(\alpha + \beta)^2}{1 + \tan(\alpha + \beta)^2} + \frac{1}{2}$$

> simplify(");

$$\cos(\alpha + \beta)^2$$

Notice that simplify made the original expression worse, but judicious use of combine and convert guided Maple to a simpler form which simplify was able to handle acceptably.

1.5   Extending the Power of Maple

Maple can be used effectively in studying physics as a symbolic-numeric-graphical calculator, without programming in the conventional sense. However, knowing how to create our own functions allows us to extend the power of Maple and make our use of it more productive. The method is illustrated below. In addition, we discuss briefly the Maple library, which has an extensive collection of specialized commands.

1.5.1   User-Defined Functions and Procedures

One of the most useful abilities of Maple is that of defining functions. There are numerous examples in this text, such as defining the position of an object as a function of time and other parameters in order to study the dependence of the motion on the various parameters of the system. Two cases of function definitions are given below, along with examples of applying those functions.

> square := (x) -> x^2;

$$square := x \rightarrow x^2$$

> square(3);   square(1+u);

$$9$$

$$(1 + u)^2$$

> X := (x0, beta, t) -> x0*(1+beta*t)*exp(-beta*t);

$$X := (x0, \beta, t) \rightarrow x0(1 + \beta t)e^{(-\beta t)}$$

> X(1, 2, . 5);   X(1, alpha, t);

$$.7357588824$$

$$(1 + \alpha t)\, e^{(-\alpha t)}$$

In these and all function definitions the variables that are given as the arguments of the function definition are local, "dummy" variables, known as *formal parameters*. They are unrelated to any other variables used in the Maple session, even those with the same name. It is possible to use globally defined variables within the body of a function definition; however, they should be used sparingly, if at all, because of the risk of unexpected side effects.

During an interactive session with Maple we sometimes find that we have generated an expression that we would like to turn into a function of certain of its parameters. We can create such a function as in the examples above, but doing so would require us to retype the expression, which could be quite long, or do some copy and paste editing. As an example, suppose we have solved a differential equation for the position of a mass moving under the influence of an ideal spring. We then wish to make a function of the solution so that we can plot the position as a function of t, for several values of initial position and velocity. We might think that this is a perfect place to use ". However, this approach will not work because the expression on the right side of -> in a function definition is not evaluated until the function is called. So if " does not work, how do we convert an expression into a function without retyping the expression? Maple has a very useful command for doing just this. The command is unapply, and we use it frequently in this text. The name arises from the following idea: a function applied to a sequence of arguments yields an expression, whereas unapply acting on an expression results in a function.

The first argument of unapply evaluates to the expression that we want to convert to a function. The subsequent arguments are variables that are found in the expression that become arguments of the function. In the following example the first line is meant to represent an expression that has been obtained from some series of previous calculations. We then call un-apply to create a function of the expression. Note that we can use " here to obtain the desired result because the arguments of the unapply command are evaluated immediately.

```
> x0*cos(omega*t) + v0/omega*sin(omega*t):
> X := unapply(", x0, v0, omega, t);
```

$$X := (x0, v0, \omega, t) \to x0 \cos{(\omega t)} + \frac{v0 \sin{(\omega t)}}{\omega}$$

```
> X(1, 0, 1, t);
```

$$\cos(t)$$

Once the X function is defined, it can be used to define a similar function for the velocity.

```
> V := (x0, v0, omega, t) -> subs(s=t, diff(X(x0,v0,omega,s), s));
```

$$V := (x0, v0, \omega, t) \to subs\left(s = t, \frac{\partial}{\partial s} X\,(x0, v0, \omega, s)\right)$$

```
> V(1,0,1,3);
```

$$- \sin(3)$$

It is necessary to use the dummy variable s in the definition so that V behaves properly when supplied with a constant value for its t argument.

Another way of defining the velocity function in the above example is

```
> V := unapply(diff(X(x0,v0,omega,t),t), x0, v0, omega, t);
```

What is the fundamental difference between these two methods? Think particularly what would be the result for V if X were subsequently redefined.

For defining more complex functions, we can use the Maple proc (procedure) construct. Its syntax, as used in this text, is as follows:

```
proc(formal parameter sequence)
    local variable sequence;
    statement 1;
    statement 2;
        . . .
    statement n;
```

end;

The formal parameter sequence can be empty if there are no arguments required by the procedure. The local line is optional. The local variable sequence should include all variables used within the procedure that are not formal arguments. This insures that they will not be considered global variables by Maple. The value returned by the procedure is that of the last statement that is executed.

In ordinary use the procedure is assigned to a variable so that it can be called by name. The example below illustrates a piecewise-defined function. It makes use of the if-then-else-fi conditional construct. The Maple programming language also includes other program flow constructs, such as while and for loops, which are not be considered in this text. See the *Maple V Language Reference Manual*[5] for more information on programming in Maple.

```
> Ramp := proc(a,x)
>     if x < a then
>         0;
>     else
>         x;
>     fi;                    # end of if statement
> end;
```
$$Ramp := proc(a,x) \text{ if } x < a \text{ then } 0 \text{ else } x \text{ fi end}$$

```
> Ramp(0, -4);   Ramp(0, 5);   Ramp(0, x);
```

$$0$$

$$5$$

Error, (in Ramp) cannot evaluate boolean

The error occurs because in the last call to Ramp the formal parameter x does not have a numerical value, so the result of the if test is unspecified.

1.5.2   Access to the Maple Library

Maple commands fall into four categories: (a) "built-in" commands, which are included in the kernel, (b) demand-loaded library commands, which are automatically loaded when called, (c) miscellaneous library commands which are not automatically loaded, and (d) related commands which are grouped together in packages. From the point of view of the average user, there is no distinction between the first two categories; any commands in these two categories are available without the user first reading them into the current session. Commands in the third category must be explicitly loaded with the readlib command (see ?readlib) prior to use. Commands in the fourth category must also be explicitly loaded. Often the with command is used to load the entire package. An alternative method is to use a "long form" of command call which consists of the package name with the command name in brackets. Examples of both of these methods of providing access to commands in the plots and linalg packages are given in later chapters.

1.6   Graphics

The graphics routines in Maple are among the most useful commands for studying physics. They fall into two major categories — two-dimensional and three-dimensional graphics. Two-dimensional graphics are used most often in this text, typically for plotting sets of functions to compare the effects of different values of the parameters in a given problem. Also commonly used is the parametric plotting facility in plotting trajectories. Specialized three-dimensional graphs include contour plots and vector field plots. Maple has facilities for each. In addition, the animation facility of Maple provides another way of looking at multivariate functions.

The basic command for plotting two-dimensional data is plot. It is used frequently enough that it is one of the commands that is automatically loaded when called. The plot command can be used to graph a single function or expression, or a set of functions and/or expressions on a single graph. It can also be used to create parametric plots, such as the trajectory of a projectile near the earth's surface, reviewed in Chapter 3. The following example illustrates simple plotting using a function of two variables.

```
> f := (beta,t) -> beta*t*exp(-beta*t):
```

We first plot $f(\beta, t)$ for the single value $\beta = 1$. The syntax is

```
> plot(f(1,t), t=0..5);
```

and the result is shown in Fig. 1.1.

Also using plot, we can show $f(\beta, t)$ as a function of t for several values of β with the command

```
> plot({f(.5,t), f(1,t), f(2,t)}, t=0..5);
```

The effect is shown in Fig. 1.2, although with color graphics output to a video screen the three curves are different colors rather than different gray levels. Because the ordering of the elements of a set are indeterminate, with this simple type of plot call we cannot control which function is assigned which color. It is possible to exert greater control by using multiple plot results as arguments to the display command. See ?display.

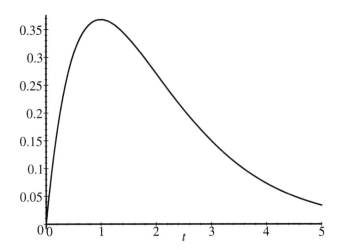

FIGURE 1.1. Plot of $\beta t\, e^{-\beta t}$ as a function of t for $\beta = 1$.

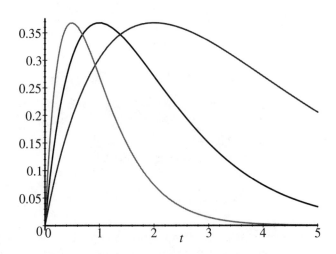

FIGURE 1.2. Plot of $\beta t\, e^{-\beta t}$ as a function of t for the three values $\beta = 0.5, 1.0, 2.0$.

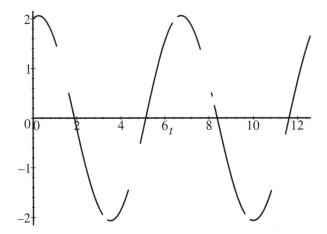

FIGURE 1.3. Illustration of gaps due to an imaginary component.

In trying to plot results obtained from the solution of physics problems there are two common instances in which an empty or incomplete plot is returned, with no explanation from Maple. The first occurs when one or more of the parameters have not been given numerical values. The second occurs when an expression is complex. Note that it is possible that an additive expression whose imaginary parts cancel can develop a small imaginary part caused by round-off errors when evaluated using floating-point arithmetic. The following example illustrates these cases.

```
> expr := (a/sqrt(a^2-1)+1)*exp(sqrt(a^2-1)*t)
>              - (a/sqrt(a^2-1)-1)*exp(-sqrt(a^2-1)*t);
```

$$expr := \left(\frac{a}{\sqrt{a^2 - 1}} + 1 \right) e^{\left(\sqrt{a^2 - 1}t\right)} - \left(\frac{a}{\sqrt{a^2 - 1}} - 1 \right) e^{\left(-\sqrt{a^2 - 1}t\right)}$$

```
> plot(expr, t=0..4*Pi);
```

The result of this plot command is a set of axes with no curve plotted. The reason is that the parameter a has not been given a numerical value. Unfortunately, Maple does not explain why the graph is blank. In the following command we supply a value for a with the limit command. We could use subs or a direct assignment to a with a similar result.

```
> plot(limit(expr, a=.25), t=0..4*Pi);
```

The resulting plot is shown in Fig. 1.3. As we see, there are gaps in the plotted curve. These are places where the use of floating-point arithmetic has caused a small imaginary part to occur, as can be seen in comparison with Fig. 1.4 which is generated by repeating the plot using only the real part.

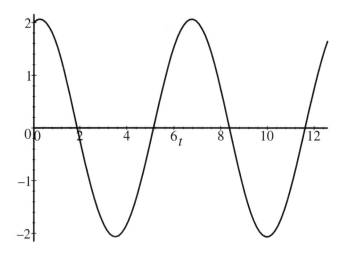

FIGURE 1.4. Plot of the real part of the floating-point expression.

> plot(Re(limit(expr,a=.25)), t=0..4*Pi);

The plot command can take a variety options to enhance the appearance of a given plot. As mentioned above, there are also a number of other kinds of plotting commands available. Information about two-dimensional parametric plotting or specialized three-dimensional plotting procedures can be obtained by entering one or more of ?plot, ?plot[options], ?plot3d, and ?plot3d[options]. In addition, the animate command can be used to help visualize expressions that depend on two or more parameters. See ?animate. There are several examples of these specialized plotting procedures in subsequent chapters of this text.

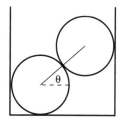

FIGURE 1.5. Problem 3.

1.7 Problems

1. Explain why Maple simply returns the input rather than evaluating the following commands:

   (a) sin(pi/2);
   (b) limit(exp(-beta*t), t=infinity);
   (c) Evalf(E);

2. Plot the Ramp(a, x) function defined in Sec. 1.5.1 for $a = 1$ over the domain $-5 \leq x \leq 5$.

3. Two logs, each of weight W, lie in a trough with vertical walls in such a way that when viewed end-on the line between their centers makes an angle θ with the horizontal. The magnitude of the forces on the bottom log due to the top log, the bottom of the trough, and one wall of the trough are F_{12}, F_{1b}, and F_{1w}, respectively. Similarly, the magnitude of the forces on the top log due to the bottom log and the other wall of the trough are F_{21} and F_{2w}. Since $F_{21} = F_{12}$, the equations for static equilibrium are

$$
\begin{aligned}
-F_{12} \cos\theta + F_{1w} &= 0 \\
F_{1b} - W - F_{12} \sin\theta &= 0 \\
F_{12} \cos\theta - F_{2w} &= 0 \\
F_{12} \sin\theta - W &= 0
\end{aligned}
$$

assuming that all frictional forces are negligible. Use Maple to

   (a) Find the force magnitudes F_{12}, F_{1b}, F_{1w}, and F_{2w};
   (b) Check the solutions;
   (c) Verify the $\theta \to \frac{\pi}{2}$ limits for the forces.

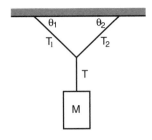

FIGURE 1.6. Problem 4.

4. A mass M is supported by three wires, as shown in the figure. The equations for static equilibrium are

$$
\begin{aligned}
T - Mg &= 0 \\
T_1 \sin \theta_1 + T_2 \sin \theta_2 - T &= 0 \\
-T_1 \cos \theta_1 + T_2 \cos \theta_2 &= 0
\end{aligned}
$$

where T, T_1, and T_2 are the tensions in the three wires.

(a) Solve the equations simultaneously to find the tensions in the wires.

(b) Show that the tensions T_1 and T_2 can be written

$$
T_1 = \frac{Mg \cos \theta_2}{\sin (\theta_1 + \theta_2)}, \quad T_2 = \frac{Mg \cos \theta_1}{\sin (\theta_1 + \theta_2)}.
$$

(c) Show that in order to support the mass when the two angles are negligibly small (*i.e.*, $\theta_1 = \theta_2 \to 0$) the tensions T_1 and T_2 must be infinitely large.

5. With Maple it is often easy to verify trigonometric identities, but not so easy to find a series of commands that will transform a trigonometric expression into an equivalent one. For example, consider the identity

$$
\sin^4 \alpha \cos^2 \beta - \sin^2 \alpha \cos^4 \beta \equiv \sin^2 \alpha \cos^2 \beta (\sin^2 \beta - \cos^2 \alpha).
$$

(a) Verify this identity by applying a single simplify or combine command to the difference between the expressions on the two sides of the equivalence.

(b) Try to reproduce the right side by starting with the expression on the left and successively applying simplify, combine, factor, *etc.* commands.

6. The electric field due to a uniformly charged circular disk of charge Q and radius R at a point on the symmetry axis a distance z from the disk center can be written as

$$\mathbf{E} = \hat{\mathbf{z}} \frac{z\sigma}{4\pi\epsilon_0} \int_0^{2\pi} d\phi \int_0^R \frac{r\,dr}{\left(z^2 + r^2\right)^{3/2}} \,,$$

where $\sigma = Q/\left(\pi R^2\right)$ is the surface charge density.

(a) Evaluate the expression for the magnitude of $\mathbf{E}$.

(b) Show that the $R \to \infty$ limit yields $\sigma/(2\epsilon_0)$, the value of the electric field due to an infinite plane of charge.

(c) Show that the lowest order term in a small z expansion is the same as the previous result, indicating that very near the disk the field is approximately that of an infinite plane.

(d) Expand the result for the magnitude of $\mathbf{E}$ as a power series in R and show that the lowest-order term is what you would expect for a point charge,

$$E = \frac{Q}{4\pi\epsilon_0 z^2} \,.$$

Review of Introductory Mechanics

2.1 Kinematics in Rectangular Coordinate Systems

Classical mechanics has traditionally been broken into three subdisciplines: kinematics, dynamics, and statics. Kinematics is the study of how motion is described in general, and thus has to be discussed before examining dynamics, the study of the laws which determine what motion actually occurs. In this text statics is treated as a special case of dynamics.

For the purposes of our discussion, a *particle* is an object which at any given instant is concentrated at a single point in space. Real objects, of course, are not particles. However, in certain contexts the finite extent of the object may be unimportant to an understanding of its motion, and the object may be considered to be a particle. Real objects can often be considered to behave as particles if their physical size is much smaller than the characteristic distances of the problem. For example, in the motion of the earth around the sun, the earth and sun may be treated as particles because their radii are much smaller than the earth-sun distance. More generally, objects may be treated as particles if interactions which depend upon the structure of the object can be neglected. For example, elastic atomic collisions, in which the atoms involved do not change their internal energy states, can be approximately treated as if the atoms are particles. Thrown baseballs can be approximated as particles if we are not concerned about the aerodynamic forces responsible for behavior such as curve balls. However, even objects whose structure cannot be ignored can often be considered to consist of a collection of particles, so that the concept of particle is generally fruitful.

The basic idea of describing the motion of a particle is to specify its location at each instant of time. This presupposes the selection of a reference system with respect to which we relate the position of the particle in question, *e.g.*, by giving a distance and a direction relative to the origin and x-axis of the reference system. The concepts of distance, direction, and time are primitives of the theory. Distance and time are measured in some standard set of units, *e.g.*, meters and seconds in the International System. Direction is determined by angular measurement (in radians, unless otherwise indicated) relative to some specified axes, or by giving distances along particular axes, or a combination of both.

This chapter is confined to kinematics using a rectangular (Cartesian) coordinate system. The position of a particle in a rectangular coordinate system is given by specifying the x-, y-, and z-coordinates of the point at which the particle is located at any given time. This position is often written as the vector $\mathbf{r}(t) = x(t)\hat{\mathbf{x}} + y(t)\hat{\mathbf{y}} + z(t)\hat{\mathbf{z}}$, where $\hat{\mathbf{x}}$, $\hat{\mathbf{y}}$, and $\hat{\mathbf{z}}$ are unit vectors in the direction of the positive x-, y-, and z-axes, respectively. Note that the unit vectors are fixed in direction and length, and are thus independent of time.

Recall that vectors are quantities which possess both a magnitude and a direction. They are represented pictorially as arrows, and are added geometrically rather than algebraically, using either the tail-to-head method or the parallelogram method. In manipulating most vectors we can move them as we wish, so long as we do not change their length or direction. Position vectors are the exception to this last property. Since they are directed from the origin to the location of the particle, their tails are anchored to the coordinate system; otherwise, they can be treated as ordinary vectors.

Of fundamental importance to the description of motion are displacement vectors, which are changes in position from one point to another. So, for example, if $\mathbf{r}_1$ and $\mathbf{r}_2$ are vectors describing the position of a particle at two different times, the displacement of the particle is the change in position, given by the vector difference $\Delta \mathbf{r} = \mathbf{r}_2 - \mathbf{r}_1$. As a particle moves through space it undergoes continuous displacement. The average *velocity* is defined as the rate of such displacement:

$$\bar{\mathbf{v}} \equiv \frac{\Delta \mathbf{r}}{\Delta t} ,$$

which in rectangular coordinates becomes

$$\bar{\mathbf{v}} = \frac{\Delta x}{\Delta t}\hat{\mathbf{x}} + \frac{\Delta y}{\Delta t}\hat{\mathbf{y}} + \frac{\Delta z}{\Delta t}\hat{\mathbf{z}} .$$

If we take the limit as $\Delta t \to 0$, we get the instantaneous velocity

$$\mathbf{v} \equiv \frac{d\mathbf{r}}{dt} . \tag{2.1}$$

In rectangular coordinates, this is

$$\mathbf{v} \equiv \frac{dx}{dt}\hat{\mathbf{x}} + \frac{dy}{dt}\hat{\mathbf{y}} + \frac{dz}{dt}\hat{\mathbf{z}} .$$

Note that the result of taking the derivative of a vector with respect to time in a rectangular coordinate system is equivalent to taking the derivative of each of the components, since the unit vectors are constant.

The rate of change of velocity is also a significant quantity. The instantaneous *acceleration* is defined to be the time derivative of the velocity:

$$\mathbf{a} \equiv \frac{d\mathbf{v}}{dt} \equiv \frac{d^2\mathbf{r}}{dt^2} . \tag{2.2}$$

Applying this to the velocity in rectangular coordinates, with v_x, v_y, and v_z being the velocity components along the three axes, we find

$$\mathbf{a} \equiv \frac{dv_x}{dt}\hat{\mathbf{x}} + \frac{dv_y}{dt}\hat{\mathbf{y}} + \frac{dv_z}{dt}\hat{\mathbf{z}} .$$

An alternate notation common in mechanics uses dots over symbols to represent time derivatives. In this notation, the definitions of the velocity and acceleration become $\mathbf{v} \equiv \dot{\mathbf{r}}$ and $\mathbf{a} \equiv \dot{\mathbf{v}} \equiv \ddot{\mathbf{r}}$.

If the position is known as a function of time, Eqs. 2.1 and 2.2 allow us to obtain the velocity and acceleration. However, often we are able to obtain the acceleration from Newton's second law; the task then is to find the velocity and position from the acceleration. To go in this reverse direction — from acceleration to velocity to position — we integrate over time. This is straightforward in rectangular coordinates since the unit vectors are constant. In general, starting with

$$\frac{d\mathbf{v}}{dt} = \mathbf{a} ,$$

we integrate both sides over time from 0 to t. Solving for $\mathbf{v}(t)$ yields

$$\mathbf{v}(t) = \mathbf{v}(0) + \int_0^t \mathbf{a}(t')\, dt' . \tag{2.3}$$

Similarly, we obtain the position as a function of time by integrating each side of Eq. 2.1. The result for $\mathbf{r}(t)$ is

$$\mathbf{r}(t) = \mathbf{r}(0) + \int_0^t \mathbf{v}(t')\, dt' . \tag{2.4}$$

If the acceleration is constant (both in magnitude and direction), Eqs. 2.3 and 2.4 are simplified to

$$\mathbf{v}(t) = \mathbf{v}(0) + \mathbf{a}t$$

and

$$\mathbf{r}(t) = \mathbf{r}(0) + \mathbf{v}(0)t + \frac{1}{2}\mathbf{a}t^2 .$$

2.2   Newton's Laws of Motion

The motion of classical objects can in principle be described with three simple laws, first stated by Sir Isaac Newton in the 17th century. His *first law* may be paraphrased as follows:

> **In the absence of outside force, a moving object continues in its state of motion in a straight line at constant speed, while an object at rest remains at rest.**

In addition to the concept of force, discussed below, implicit within this statement is the existence of a reference system with respect to which the object is moving at constant velocity, including zero as a special case. For our purposes, we assume that reference systems do exist in which this law is valid; such systems are *inertial reference systems*. If one inertial reference system can be found, so can others: any reference system moving at a constant velocity with respect to an inertial reference system is itself inertial.

In actuality, all objects interact with their surroundings, and so are subject to outside forces. Newton's second and third laws directly deal with outside forces. The *second law* may be stated briefly as follows:

> **The magnitude of the acceleration of an object acted upon by a force is directly proportional to the magnitude of the force, and inversely proportional to the mass of the object. The direction of the acceleration is the same as that of the force.**

Written as an equation, this is

$$\mathbf{F} = m\mathbf{a} , \tag{2.5}$$

where $\mathbf{F}$ is the force acting on the object. The mass in this law, more appropriately termed the inertial mass, is a measure of a given object's resistance to a change in its motion. It is usually defined operationally in such a way that the mass of a given object is determined by comparing its acceleration to that of a standard mass, when both are subjected to the same magnitude force.

If the mass of the object under consideration is constant, the second law can be written in the form

$$\mathbf{F} = \frac{d(m\mathbf{v})}{dt} = \frac{d\mathbf{p}}{dt} ; \tag{2.6}$$

the product of the mass and velocity defines the *momentum* of a particle. Thus, the net force is equal to the time rate of change of the momentum. This is actually the form in which Newton proposed it, and is correct for particles moving near the speed of light, where Eq. 2.5 is not valid. This text considers the realm in which Eqs. 2.5 and 2.6 are nearly equivalent, and either may be referred to as Newton's second law.

In each of Newton's laws of motion, the two given above as well as the third which remains to be stated, the concept of *force* plays a key role. There has been, and continues to be, significant debate about a definition for force. Refer to the classical mechanics texts listed in the references for details. Since our interest is not in the formal analysis of the foundations of classical mechanics, we simply assume that forces are given by separate physical laws which specify functional forms for, *e.g.*, the gravitational or electromagnetic interactions. Newton's second law is an empirically-based statement that such individual forces add vectorially to give a *net force* which determines the acceleration of an object by Eq. 2.5 or 2.6.

Newton's *third law* may be stated as follows:

> **Whenever one object exerts a force on a second object, the second object exerts a force of equal magnitude and opposite direction upon the first object.**

Although the third law has been empirically found to hold under a wide range of conditions, there are circumstances for which it is violated. For example, the law does not in general hold true for forces which propagate at a finite speed between the two objects. Fortunately, most of classical mechanics does not rest upon the validity of the third law. In addition, for a very wide range of phenomena the third law is approximately valid.

There are other issues, some rather subtle, that are raised by Newton's laws of motion. Newton himself was aware of some of them, while others escaped notice for over two hundred years. A number of scientists (notably Euler, Mach, and Einstein) have carefully examined the assumptions implicit in the laws of mechanics and placed the science on firm theoretical ground. See [11] for a more recent systematic examination of the foundations of mechanics.

2.3   Examples of Motion Under Constant Forces

Section 2.1 discussed kinematics and the general problem of obtaining the position and velocity of a particle-like object from knowledge of its acceleration. Newton's second law relates the acceleration of an object to the forces acting on it. Thus, in principle, if we can find all the forces on one or more objects as functions of time, we can use Eq. 2.5 to find the acceleration(s) and then solve for the position and velocity of the object(s) with Eqs. 2.3 and 2.4. The difficulty with this prescription is that physics usually gives expressions for the net force as a function of position and/or velocity rather than explicitly as a function of time. An important exception to this usual circumstance is the case in which all forces are constant in time. This section considers examples of motion of objects subject to constant forces.

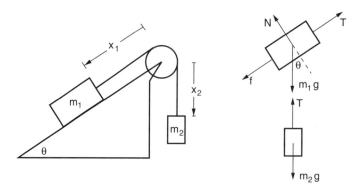

FIGURE 2.1. Schematic of and force diagrams for the pulley and incline system.

There are two major goals for studying these examples. The primary goal of these and all examples in this text is to provide help in understanding physics. The examples of this section (or similar ones) are covered in introductory physics courses, so the physics they illustrate should be familiar. If not, use these examples and the problems at the end of the chapter to bring your expertise up to the level that they represent.

The second major goal is to learn how to use Maple as a tool for solving problems and elucidating the physics that they contain. Maple is used throughout the text to analyze motion of a variety of physical systems. It is important to become comfortable with the basic operations and commands of the system at this point, where the physics is familiar. Keep in mind, however, that in this text Maple is presented as a *tool for investigating physics*. If it is easier to do some simplification by hand as opposed to trying to cajole Maple to yield a particular form, do so.

2.3.1 *Pulley and Incline System*

Figure 2.1 shows a system of two masses, one (m_1) on an incline, and the other (m_2) suspended by a rope. The rope passes over a frictionless pulley and is fastened to m_1. Let us assume that m_2 falls, so that m_1 is pulled up the incline, and that the coefficient of friction between m_1 and the incline is μ. We analyze the forces and the resulting motion using the coordinates x_1 and x_2 shown in the figure, with positive directions as indicated by the arrows. This is equivalent to setting up individual Cartesian coordinate systems for each mass, with a common origin at the center of the pulley, but differently directed x-axes.

Figure 2.1 also shows the force diagrams for the two masses. T represents the tension in the rope, and N the normal force between the incline and m_1. The kinetic friction is given by $f = \mu N$. We let a_1 be the acceleration of m_1 in the direction of increasing x_1 (down the incline), and a_2 the acceleration of m_2 in the direction of increasing x_2 (downwards). Newton's second law applied to m_1 yields the following two equations, for directions along the incline and perpendicular to it.

$$\mu N + m_1 g \sin \theta - T = m_1 a_1 \qquad (2.7)$$

$$N - m_1 g \cos \theta = 0 . \qquad (2.8)$$

Similarly, for m_2,

$$m_2 g - T = m_2 a_2 . \qquad (2.9)$$

In addition, if the rope does not stretch, there is the constraint that

$$x_1 + x_2 = constant .$$

Differentiating this constraint twice with respect to time results in

$$a_1 + a_2 = 0 . \qquad (2.10)$$

Equations 2.7 through 2.10 represent a set of four algebraic equations that can be solved for four unknowns, a_1, a_2, T, and N. This can be done by hand, but the algebra is tedious, and the likelihood of error is high. In applying Maple to the problem, we give each equation a label and, for convenience in checking the solution, transpose all terms to the left side of the equal sign. The following are the four equations as they are entered into Maple.

```
> restart;
> eq1 := mu*N + m1*g*sin(theta) - T - m1*a1 = 0:
> eq2 := N - m1*g*cos(theta) = 0:
> eq3 := m2*g - T - m2*a2 = 0:
> eq4 := a1 + a2 = 0:
```

These equations can be solved with Maple's solve command. The set of equations forms the first argument, and the set of unknowns the second argument. For convenience, the resulting solution set is assigned to a variable name.

```
> sol := solve({eq1,eq2,eq3,eq4}, {a1,a2,T,N});
```

The result is returned in a form similar to the following:

$$sol := \{a2 = ..., T = ..., N = ..., a1 = ...\} .$$

The ordering of the elements in the solution set is internally determined by Maple. With some manual rearrangement, the solution is

$$a_1 = -a_2 = \frac{(m_1 \sin \theta + \mu m_1 \cos \theta - m_2)g}{m_1 + m_2} ,$$

$$T = \frac{m_1 m_2 (1 + \sin\theta + \mu\cos\theta)g}{m_1 + m_2} ,$$

and

$$N = m_1 g \cos\theta .$$

We check the Maple solution by substituting back into the original equations with the subs command. Applying simplify to the results of subs makes it easier to verify the solution. Thus the Maple command

> simplify(subs(sol,{eq1,eq2,eq3,eq4}));

yields

$$\{0 = 0\} .$$

This simple result occurs because the original equations were written with zeros on the right sides. If this had not been done, the results of the solution check would have been a set of equations whose left and right sides have to be visually verified as being equal. Remember that since the result is a set, the four identical $0 = 0$ elements have been reduced to one.

At this point Maple has not assigned any values to the variables a1, a2, T, or N. We do so with the assign command.

> assign(sol);

After this command is executed each of the unknowns has the value given on the right side of the equal sign in the corresponding equation of the solution set.

In checking a solution to a physics problem, whether the solution is obtained by hand, by numerical evaluation in a specially written computer program, or with a problem solving software package like Maple, it is extremely important to check the results by verifying that they agree with limiting cases for which the answer is known or can be inferred from the physics of the problem. In the present context, with the assignment of the unknowns, we can conveniently verify a variety of limiting cases. For example, if $m_1 = 0$, the suspended mass should accelerate downward at the rate $a_2 = g$. We check this with the limit function.

> limit(a2, m1=0);

$$g$$

Similarly, if $m_2 = 0$ and friction is negligible ($\mu = 0$), the mass on the incline will experience an acceleration which is directed down the plane at a rate $a_1 = g\sin\theta$. This is easily verified by Maple with

> limit(limit(a1, m2=0), mu=0);

$$g \sin(\theta)$$

There are other obvious limiting cases that can be checked.

As an exercise, examine the $\theta = 0$ and $\theta = \frac{\pi}{2}$ limits and satisfy yourself that they are correct.

Examining limiting cases of general expressions provides important checks on answers to problems. It also reveals much of the physics inherent in a given problem by allowing us to focus on particular aspects while neglecting complicating factors. However, we must be careful not to choose limiting cases which lie outside the domain of assumptions that were made in solving a given problem. For example, in the above solution we might look at the $\mu \to \infty$ limit of a_2, reasoning that if the friction on m_1 were infinitely large, m_1 will not slide up or down the incline, so that both a_1 and a_2 go to zero. If we check the limit, however, we find that $a_1 \to \infty$ and $a_2 \to -\infty$! More careful thought indicates that this result is correct because when we included friction in the parallel-to-the-incline equation of motion for m_1, we assumed that the mass was moving up the plane. Had it been at rest, we would have had to use static friction, which is bounded by μN, but generally not equal to it. The infinite accelerations in the $\mu \to \infty$ limit tell us that the assumed motion will stop instantly.

2.3.2 Mass Sliding Down a Movable Incline

The system in this example consists of a mass m which is placed on the incline of a wedge of mass M. The wedge rests on a table. There is no friction between any pair of surfaces. The system is shown in Fig. 2.2, along with force diagrams for the two masses. The force T is the normal force exerted by the table, and N is the normal force between the wedge and the mass m. The coordinates we use are also shown in the figure. They represent the horizontal and vertical position components for the mass m, and the horizontal position of the wedge.

Applying Newton's second law to the x- and y-motion of m, and the X-motion of M results in the following equations.

$$-N \sin \theta = ma_x \tag{2.11}$$

$$N \cos \theta - mg = ma_y \tag{2.12}$$

$$N \sin \theta = MA_x . \tag{2.13}$$

If we wished to find T, we could include the vertical equation for M.

We see from the figure that the chosen coordinates denote the position of the two masses with respect to an inertial coordinate system, so that Newton's laws apply. Were we, for example, to have taken coordinates for

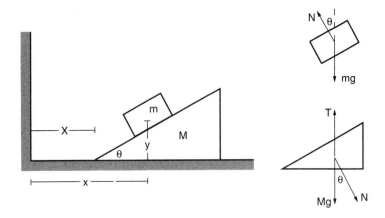

FIGURE 2.2. Schematic of and force diagrams for a mass sliding down a moveable incline.

mass m relative to some point fixed on the wedge, we could not use Eq. 2.5 to get equations of motion for m because such a system is non-inertial.

In addition to Eqs. 2.11 through 2.13, the fact that mass m slides along the plane of incline means that there is a constraint relating the three coordinates x, y, and X. In particular, from the figure,

$$\tan\theta = \frac{y}{x - X}\ .$$

We neglect the size of m in writing this constraint. Rearranging this equation and differentiating twice with respect to time leads to our fourth equation,

$$(a_x - A_x)\tan\theta = a_y\ . \tag{2.14}$$

To solve these four numbered equations and check the solution with Maple, we proceed with the following sequence of commands.

```
> restart;
> eq1 := -N*sin(theta) - m*ax = 0:
> eq2 := N*cos(theta) - m*g - m*ay = 0:
> eq3 := N*sin(theta) - M*Ax = 0:
> eq4 := (ax - Ax)*tan(theta) - ay = 0:
> sol := simplify(solve({eq1,eq2,eq3,eq4}, {ax,ay,Ax,N}));
```

$$sol := \left\{ ax = \frac{\cos(\theta)Mg\sin(\theta)}{\%1},\ N = -\frac{\cos(\theta)mMg}{\%1} \right.$$

$$\left. ay = -\frac{g\left(-M + \cos(\theta)^2M - m + m\cos(\theta)^2\right)}{\%1},\ Ax = -\frac{\cos(\theta)mg\sin(\theta)}{\%1} \right\}$$

$$\%1 := -M - m + m\cos(\theta)^2$$

> simplify(subs(sol,{eq1,eq2,eq3,eq4}));

$$\{0 = 0\}$$

The $\{0 = 0\}$ which is returned gives confidence in the solution.

Two obvious limiting cases to check are $\theta = 0$ and $\theta = \frac{\pi}{2}$. In the former case, all accelerations are zero and the normal force N is equal to the weight of mass m; in the latter case, m accelerates downward with the acceleration due to gravity ($a_y = -g$) and all other unknowns are zero. We verify these cases by substituting appropriate values for θ in the expression for sol. That is,

> simplify(subs(theta=0, sol));

$$\{ax = 0, \ Ax = 0, \ ay = 0, \ N = mg\}$$

> simplify(subs(theta=Pi/2, sol));

$$\{ax = 0, \ Ax = 0, \ N = 0, \ ay = -g\}$$

To further investigate the solution to this problem we assign values to the unknowns.

> assign(sol);

Now let us look at the limiting case of a mass sliding on a fixed incline. We obtain it by using Maple to evaluate the $M \to \infty$ limit for the magnitude of the acceleration of mass m:

> limit(sqrt(ax^2+ay^2), M=infinity);

$$\sqrt{g^2 \left(\cos(\theta)^2 \sin(\theta)^2 + 1 - 2\cos(\theta)^2 + \cos(\theta)^4\right)}$$

> simplify(");

$$\sqrt{-g^2 \cos(\theta)^2 + g^2}$$

Maple preferentially uses $\cos^2$ rather than $\sin^2$ functions in its simplifications. Here that is obviously a disadvantage. To get Maple to put this into a more acceptable form requires more experience with the program. Perhaps the easiest thing to do in such a circumstance is to finish the result by hand, using the identity $1 - \cos^2 \theta \equiv \sin^2 \theta$ and selecting the root that gives an overall positive result since we are examining the *magnitude* of the acceleration.

You are encouraged to experiment with Maple to gain the experience that will enable you to use the program most effectively.

However, keep in mind that your goal is to learn physics. If you find that it is easier to complete a calculation by hand, do so.

With the aforementioned caveat in mind, the solution to this problem can be summarized with the following equations:

$$a_x = -\frac{Mg \sin(2\theta)}{2(M + m \sin^2 \theta)} \tag{2.15}$$

$$a_y = -\frac{(M + m) \sin^2 \theta}{M + m \sin^2 \theta} \tag{2.16}$$

$$A_x = \frac{mg \sin(2\theta)}{2(M + m \sin^2 \theta)} \tag{2.17}$$

$$N = \frac{Mmg \cos \theta}{M + m \sin^2 \theta} \, . \tag{2.18}$$

We note that $MA_x = -ma_x$, a consequence of Newton's third law. This can be verified with the Maple statement

> simplify(m*ax+M*Ax);

                    0

which yields 0 since ax and Ax were previously assigned their appropriate values.

Repeat the calculations with a horizontal external force F applied to the wedge. How large and in what direction must F be to cause m to remain fixed with respect to the wedge? Hint: One way of finding the answer for the magnitude is to use Maple to find the value of F that satisfies the equation $a_y = 0$. Will $MA_x = -ma_x$ in this case? Explain.

2.3.3 Force Applied to Stacked Boxes

Figure 2.3 illustrates a third example of applying Newton's laws. The system consists of two masses, with m_2 sitting on m_1, which is in turn on a table. A force **F** is applied to m_2 at an angle θ with respect to the horizontal. We assume that m_1 and m_2 are moving to the right, that the coefficient of kinetic friction between m_1 and the table is μ, and that the coefficient of static friction between the two masses is sufficiently large that they do not slide with respect to each other. We will (a) show that there is a critical angle

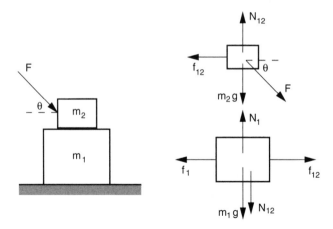

FIGURE 2.3. Schematic of and force diagrams for stacked boxes subject to an applied force.

θ_c such that if $\theta \geq \theta_c$, the force **F** cannot maintain the motion of the system toward the right regardless of its size, (b) find the value of F for which the two masses move with constant velocity, assuming $\theta < \theta_c$, and (c) determine the requirement for the coefficient of static friction between m_1 and m_2 when the masses are moving with constant velocity.

Employing the force diagrams in Fig. 2.3, we find that Newton's second law gives the following four equations of motion; the positive x-direction is horizontal and to the right.

$$f_{12} - f_1 = m_1 a_{1x} ,\qquad(2.19)$$

$$N_1 - N_{12} - m_1 g = 0 ,\qquad(2.20)$$

$$F \cos\theta - f_{12} = m_2 a_{2x} ,\qquad(2.21)$$

and

$$N_{12} - F \sin\theta - m_2 g = 0 .\qquad(2.22)$$

These are entered into Maple with

```
> restart;
> eq1 := f12 - f1 - m1*a1x = 0:
> eq2 := N1 - N12 - m1*g = 0:
> eq3 := F*cos(theta) - f12 - m2*a2x = 0:
> eq4 := N12 - F*sin(theta) - m2*g = 0:
```

This example considers only the case for no relative sliding between m_1 and m_2, so we take $a_{1x} = a_{2x} = a$. In addition, m_1 is sliding along the table, so that $f_1 = \mu N_1$. Hence, after we add the assignments

```
> a1x := a:    a2x := a:    f1 := mu*N1:
```

we have Maple solve the four equations, check the solution, and assign values to the unknowns with

```
> sol := solve({eq1,eq2,eq3,eq4}, {a,N1,N12,f12}):
> simplify(subs(sol,{eq1,eq2,eq3,eq4}));
```

$$\{0 = 0\}$$

```
> assign(sol);
```

We verify the two limiting cases of no applied force and no friction between m_1 and the table. If there is no applied force, only the sliding friction acts on the combination of m_1 and m_2 in the x-direction, so $a \to -\mu g$.

```
> limit(a, F=0);
```

$$-\mu g$$

On the other hand, if there is no sliding friction, the combination moves in response to the x-component of the applied force only.

```
> limit(a, mu=0);
```

$$\frac{F \cos(\theta)}{m1 + m2}$$

Other limiting cases may be similarly investigated.

Examining the acceleration as a function of the angle and magnitude of F further aids in our understanding of the problem. To do this, we define a function which gives the acceleration as a function of F and θ,

```
> A := unapply(a, F, theta):
```

and choose some typical values (in SI units) for the parameters. For example,

```
> m1 := 3:   m2 := 1:   mu := .8:   g := 9.8:
```

Plotting $A(F,\theta)$ for $0 \le \theta \le \frac{\pi}{2}$ and three different values of F,

```
> plot({A(10,theta), A(100,theta), A(10000,theta)}, theta=0..Pi/2, -20..50);
```

returns the curves shown in Fig. 2.4. The third argument of this plot command, which is optional, restricts the vertical range plotted. As we see, for small values of F the acceleration is negative for any θ because F is unable to overcome the friction acting on the lower box. For smaller angles, larger values of F overcome friction to produce a positive acceleration. As θ increases, however, the horizontal component of F decreases, while the frictional force

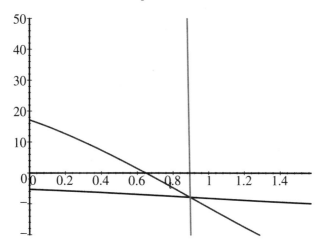

FIGURE 2.4. Acceleration of the stacked boxes as a function of the angle of the applied force, for three values of the force magnitude.

increases (due to the increasing vertical component of **F**) so that the acceleration again becomes negative. The figure also suggests that there is some limiting angle beyond which the acceleration is negative for any magnitude of **F**. The following procedure demonstrates that this is the case.

We unassign the parameters that were given values and look for the magnitude of the applied force **F** which causes the acceleration to vanish.

```
> m1 := 'm1':   m2 := 'm2':   mu := 'mu':   g := 'g':
> solve(a=0, F);
```

$$-\frac{\mu\, m1\, g + \mu\, m2\, g}{-\cos(\theta) + \mu\, \sin(\theta)}$$

Let us call this magnitude F_0:

```
> Fo := factor(");
```

so that the answer to part (b) is

$$F_0 = \frac{\mu g (m_1 + m_2)}{\cos\theta - \mu\sin\theta}, \tag{2.23}$$

provided that it is positive. We see that this requires that θ be less than some critical angle θ_c at which the denominator of the right side vanishes. This can be found by hand, or Maple can do it with

```
> theta[c] := solve(denom(Fo)=0, theta);
```

(The [c] provides the subscript on theta.) The result is

$$\theta_c = \arctan(\frac{1}{\mu}) \, . \tag{2.24}$$

We note that as the table surface becomes frictionless ($\mu \to 0$), $\theta_c \to \frac{\pi}{2}$. In this limit any force $\mathbf{F}$ which has a positive x-component will cause the two masses to accelerate to the right.

Looking back at Fig. 2.4, it appears that all the $A(F, \theta)$ curves intersect at a particular value of θ. This is indeed the case, not only for the parameter values of this figure, but for any set of parameters. We can verify this by choosing two arbitrary values for F and solving for the angle at which the corresponding accelerations are equal.

> solve(A(F1,theta)=A(F2,theta), theta);

$$\arctan\left(\frac{1}{\mu}\right)$$

The result is the critical angle as given in Eq. 2.24. Substituting this value for θ into the expression for the acceleration and simplifying

> simplify(A(F,theta[c]));

we find that

$$a(\theta = \theta_c) = -\mu g \, .$$

This is just the value of the acceleration when $F = 0$, as it should be since at this angle the acceleration is the same for all values of F. Thus, another interpretation of the critical angle is that angle at which the boxes move as if there were no applied force $\mathbf{F}$. Physically, θ_c is the angle at which the additional frictional force due to the larger normal force (caused by the vertical component of $\mathbf{F}$) is exactly cancelled by the horizontal component of $\mathbf{F}$.

Finally, the static frictional force between m_1 and m_2 is related to the normal force between the two masses by the inequality $f_{12} \leq \mu_{12} N_{12}$, where μ_{12} is the coefficient of static friction between the two boxes. Hence we find the minimum value for μ_{12} when the two masses are moving with constant velocity to the right by taking $F = F_0$ and evaluating the ratio f_{12}/N_{12}. From the Maple command

> limit(f12/N12, F=Fo);

we obtain the following result for part (c):

$$\mu_{12} \geq \frac{\mu(m_1 + m_2)\cos\theta}{\mu m_1 \sin\theta + m_2 \cos\theta} \, . \tag{2.25}$$

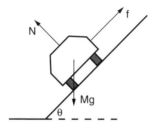

FIGURE 2.5. Schematic of a car moving on a banked curve.

You should expect that in the $\mu = 0$ limit the force F_0 required to maintain the motion would be zero, and no friction between m_1 and m_2 would be required; *i.e.*, $f_{12} = 0$. Verify these expectations.

2.3.4   Car Moving Around a Banked Curve

As our final example of motion in the presence of constant forces, let us consider a car of mass M moving at a constant speed v around a circular curve of radius R. The roadway is banked, making an angle θ with the horizontal, as seen in the cross-sectional view shown in Fig. 2.5. Strictly speaking, this is not an example of constant forces since the directions of the forces change with time. However, since the magnitudes are constant, and we treat the radial and tangential components independently, the mathematics is similar to the other problems that we have examined in this section.

An object moving in uniform circular motion undergoes an acceleration directed toward the center of the circle. The magnitude of the acceleration is given by

$$a_c = \frac{v^2}{R},$$

where v and R are respectively the speed of the object and the radius of its circular path. This *centripetal acceleration* is a special case of the more general expression for acceleration in polar coordinates, which is found in Sec. 3.1.1.

We apply Newton's 2nd law to this problem by equating the horizontal component of the net force (directed toward the center of the circle) to Ma_c, and setting the vertical component of the net force equal to zero. In the equations of motion below we assume that the static frictional force between the tires and the road is directed up the incline, as indicated in the figure.

Depending on the values of the parameters of the problem, however, this force could instead be down the plane. If so, the solution will yield a negative value for f, as we will see.

The equations of motion are

$$N \sin \theta - f \cos \theta = M \frac{v^2}{R}$$

and

$$N \cos \theta + f \sin \theta - Mg = 0 .$$

They are solved, checked, and assigned with the following sequence of commands:

```
> restart;
> eq1 := N*sin(theta) - f*cos(theta) = M*v^2/R:
> eq2 := N*cos(theta) + f*sin(theta) - M*g = 0:
> sol := solve({eq1,eq2}, {f,N}):
> simplify(subs(sol,{eq1,eq2}));
```

$$\left\{ 0 = 0, \frac{Mv^2}{R} = \frac{Mv^2}{R} \right\}$$

```
> simplify(sol);
```

$$\left\{ N = \frac{M \left(\sin(\theta)v^2 + gR \cos(\theta) \right)}{R}, f = \frac{M \left(g \sin(\theta)R - v^2 \cos(\theta) \right)}{R} \right\}$$

```
> assign(");
```

The normal force is of course positive for any physically reasonable value for the parameters, but the frictional force can be either positive or negative. If $v > \sqrt{gR \tan \theta}$, $f < 0$, indicating that the frictional force is directed down the plane. This is because for higher speeds the horizontal component of the normal force is insufficient to compel the car to move in its circular path. For low speeds, $v < \sqrt{gR \tan \theta}$, friction is directed up the plane to keep the car from sliding down. There are, of course, limitations to f imposed by the coefficient of friction between the tires and the road; we assume that f is within these limitations in considering uniform circular motion.

The speed $v = \sqrt{gR \tan \theta}$ is the critical speed for which the banking of the curve was designed. It is obtained by looking for the speed at which there is no friction between the tires and the road. Since we assigned a value to f earlier, this is easily found with the Maple command

```
> solve(f=0, v);
```

$$-\frac{\sqrt{\cos(\theta)g \sin(\theta)R}}{\cos(\theta)}, \frac{\sqrt{\cos(\theta)g \sin(\theta)R}}{\cos(\theta)}$$

Since we seek the magnitude of the velocity, the second root is the desired one. Note, however, that Maple is not easily coerced into further simplifying the result by combining the $\cos\theta$ factors under the radical. After a few obvious tries such as simplify(...) or combine(..., power) without success, it is perhaps best to complete the simplification manually.

2.4   Conservation of Mechanical Energy

There are many problems that can be solved more easily by using energy concepts than by applying Newton's laws of motion. The solution of such problems rests on two theorems whose proofs, though simple, are not presented here. The first is the *work-energy theorem*, which states that the work done by the net force on a particle as it moves from one point to another is equal to the change in the particle's kinetic energy over that motion:

$$\int_{\mathbf{r}_a}^{\mathbf{r}_b} \mathbf{F}_{net} \cdot d\mathbf{r} = \frac{1}{2}mv_b^2 - \frac{1}{2}mv_a^2 \ .$$

If the only forces that do work on a particle are conservative forces (*i.e.*, derivable from a potential energy function), the work-energy theorem can be used to prove that the *mechanical energy* of the particle, the sum of its kinetic and potential energies, remains constant throughout its motion. That is, when only conservative forces do work, $K + U$ is constant, where K and U are the kinetic and potential energies, respectively. For any two points, a and b, an alternative way of expressing this is

$$K_a + U_a = K_b + U_b \ .$$

The two problems below are examples of the application of this result.

2.4.1   Motion of a Bead on a Wire

Let us consider a vertically-oriented circular wire loop of radius R, around which a bead of mass m moves under the influence of gravity. Friction is negligible. The bead is located by specifying θ, its angular position from the top of loop, as seen in Fig. 2.6. We assume that the bead starts at rest at $\theta = 0$, and analyze its motion. (Actually, since the top of the loop is an unstable equilibrium position, it is necessary to give the bead a small shove, or start it slightly off the top position in order for it to move.)

There are only two forces acting on the bead — that due to gravity, which is downward in the figure, and that due to the wire. With no friction, this latter force is directed normal to the wire, and thus can do no work. Hence, the following equation describes conservation of energy, with the zero potential energy level passing through the center of the loop:

$$mgR = \frac{1}{2}mv^2 + mgR\cos\theta \ . \tag{2.26}$$

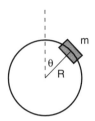

FIGURE 2.6. Schematic of a bead moving around a circular wire loop.

Thus, the energy conservation equation is sufficient to find the speed of the particle as a function of its position.

If, however, we wish to determine the force that the wire exerts on the bead at each position as well, we can appeal to Newton's 2nd law as it relates to centripetal force and acceleration. Choosing the positive direction away from the center yields

$$-mg\cos\theta + N = -\frac{mv^2}{R}, \tag{2.27}$$

where $mg\cos\theta$ is the component of the weight directed toward the center of the wire loop, and N is the force that the wire exerts on the bead.

The Maple solution of the problem is accomplished with the following statements.

```
> restart;
> eq1 := m*g*R - 1/2*m*v^2 - m*g*R*cos(theta) = 0:
> eq2 := -m*g*cos(theta) + N + m*v^2/R = 0:
> sol := solve({eq1,eq2}, {N,v});
```

$$sol := \{N = 3mg\cos(\theta) - 2mg, \; v = RootOf\left(-2gR + \_Z^2 + 2gR\cos(\theta)\right)\}$$

The RootOf expression in the result is Maple's shorthand way of referring to multiple possible solutions. (See ?RootOf.) These occur in this problem because the equations are quadratic in the unknown v. Maple can be forced to give all the solutions explicitly with the allvalues command. Thus,

```
> allvalues(sol);
```

$$sol := \left\{ v = -\sqrt{2gR - 2gR\cos(\theta)}, \; N = 3mg\cos(\theta) - 2mg \right\},$$
$$\left\{ N = 3mg\cos(\theta) - 2mg, \; v = \sqrt{2gR - 2gR\cos(\theta)} \right\}$$

There are two sets of solutions, corresponding to positive and negative values for v. Since our concern is with the speed, we choose the positive value and assign values accordingly.

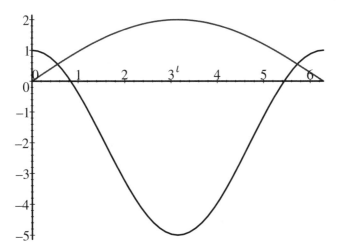

FIGURE 2.7. The normal force and the speed as a function of the angular position of the bead on the wire loop.

> assign("[2]);

Note that the ordering of the solutions by Maple can vary, so that "[1] may be required in a particular session instead of "[2]. The results for v and N are thus

$$v = \sqrt{2gR(1 - \cos\theta)} \tag{2.28}$$

and

$$N = mg(3\cos\theta - 2) . \tag{2.29}$$

At the top of the loop ($\theta = 0$), $v = 0$ and $N = mg$. N thus cancels the gravitational force, as it should, since the centripetal acceleration is zero where the bead is at rest. In addition, Eq. 2.28 expresses the expected result that the speed of the particle at a given location is proportional to the square root of the vertical distance that it has fallen.

Unitless forms of N and v as functions of θ are plotted with the statement

> plot({N/(m*g), v/sqrt(g*R)}, theta=0..2*Pi);

The result is shown in Fig. 2.7.

As can be seen from the plot, as well as from Eq. 2.29, the normal force due to the wire changes from positive (outward) to negative (inward) as it passes through the angle $\theta = \arccos(2/3)$. If a mass were sliding without friction on a spherical or cylindrical surface, rather than a bead tracking along a wire, this is the angle at which the mass would lose contact with the surface. The gravitational force would no longer be able to supply the required centripetal force for angles greater than this because of the combination of decreasing centripetal component and increasing speed. The wire, however, is able to

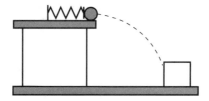

FIGURE 2.8. Schematic of a ball projected by a spring cannon.

pull in on the bead to keep it in its circular motion. Figure 2.7 clearly shows, as expected, that the force due to the wire has its greatest magnitude at the point where the bead's speed is maximum: *i.e.*, at the bottom of the loop.

The connection between N and v can be made explicit by using Maple's parametric plotting capabilities to graph N vs. v (although for this problem the parametric plotting feature is not necessary since Eqs. 2.26 and 2.27 can be used to obtain an explicit expression for N as a function of v). It is perhaps more revealing, however, to plot N vs. v^2. Try it.

2.4.2   A Spring-Powered Cannon

A physics professor is using a spring at the edge of a table to launch a ball horizontally in an attempt to land it in a waste basket. On the first attempt she compresses the spring by 1.0 cm, with the result that the ball is short of its target by 40 cm. If the horizontal distance to the near edge of the waste basket is 2.0 m, by how much should she compress the spring for her second attempt so that the ball just clears the near edge of the basket?

The situation is shown in Fig. 2.8. We take the origin of the coordinate system to be the point where the ball loses contact with the spring, and the positive x- and y-directions to be to the right and downward, respectively. We let h be the vertical distance from the table top to the top of the waste basket. Since $v_{0y} = 0$, the time required for the ball to fall the distance h is determined from the kinematic equation

$$h = \frac{1}{2}gt_f^2 \, . \tag{2.30}$$

It is independent of the horizontal component of the ball's velocity. The horizontal distance travelled during the time t_f is given by

$$d = v_x \, t_f \, . \tag{2.31}$$

Finally, since the spring force is conservative, conservation of energy relates the speed at which the ball leaves the spring to the amount, x, that the spring is compressed,

$$\frac{1}{2}kx^2 = \frac{1}{2}mv_{0x}^2 .$$ (2.32)

Since there are no horizontal forces acting on the ball during its flight, $v_x = v_{0x}$, and we have a sufficient number of equations to solve for the relationship between d and x. We first enter Eqs. 2.30 through 2.32 into Maple, using the symbols v and t to represent v_x and t_f:

```
> restart;
> eq1 := h = 1/2*g*t^2:
> eq2 := d = v*t:
> eq3 := 1/2*k*x^2 = 1/2*m*v^2:
```

Then we have Maple solve them with the command

```
> sol := solve({eq1,eq2,eq3}, {d,v,t});
```

$$sol := \{t = RootOf\left(-2h + g\_Z^2\right), v = RootOf\left(kx^2 - m\_Z^2\right),$$
$$d = RootOf\left(kx^2 - m\_Z^2\right) RootOf\left(-2h + g\_Z^2\right)\}$$

As mentioned earlier, Maple can be forced to expand the RootOf expressions into explicit solutions with the allvalues command. It is important to use the 'd' option here so that the same roots from the RootOf expressions in the t and v results are used in the result for d. (See ?RootOf.)

```
> allvalues(sol, 'd');
```

$$\left\{v = \frac{x\sqrt{mk}}{m}, d = \%1, t = \frac{\sqrt{2}\sqrt{gh}}{g}\right\} \quad \left\{v = -\frac{x\sqrt{mk}}{m}, t = \frac{\sqrt{2}\sqrt{gh}}{g}, d = -\%1\right\}$$

$$\left\{v = \frac{x\sqrt{mk}}{m}, d = -\%1, t = -\frac{\sqrt{2}\sqrt{gh}}{g}\right\} \quad \left\{v = -\frac{x\sqrt{mk}}{m}, d = \%1, t = -\frac{\sqrt{2}\sqrt{gh}}{g}\right\}$$

$$\%1 := \frac{x\sqrt{mk}\sqrt{2}\sqrt{gh}}{mg}$$

Of these four possible solution sets, only the one in which x, v, and t are all positive is physically acceptable. It is selected with

```
> "[1];
```

The current version of Maple is rigorous in its mathematics, and will not simplify the result unless it is told the quantities m, k, h, and g are positive (and hence real). Although we could give it this information with assume commands, it is easy to apply manual simplification at this point. The solution becomes

$$\left\{ v = \sqrt{\frac{k}{m}}x, \ t = \sqrt{\frac{2h}{g}}, \ d = \sqrt{\frac{2hk}{mg}}x \right\} . \tag{2.33}$$

These results can verified as solutions to the equations in the usual way. Furthermore, a quick examination of Eq. 2.33 verifies that the Maple solution has the correct $x = 0$, $k = 0$, $h = 0$, $g = 0$, and $m \to \infty$ limits.

For problems with multiple solution sets, it is often easier to use a one-equation-at-a-time approach to eliminating non-physical roots rather than treating the equations simultaneously. That is, we solve for an unknown in one equation (selecting the appropriate root, if necessary, based on the physics) and substitute the result into the other equations. We then repeat the procedure until all equations have been treated. The operation of Maple makes it especially easy to take this approach. For example, we can solve the above problem in the manner given below.

```
> solve(eq1, t);
```

$$-\frac{\sqrt{2}\sqrt{gh}}{g}, \ \frac{\sqrt{2}\sqrt{gh}}{g}$$

```
> t := "[2]:
> solve(eq3, v);
```

$$-\frac{x\sqrt{mk}}{m}, \ \frac{x\sqrt{mk}}{m}$$

```
> v := "[2]:
> d := solve(eq2, d);
```

$$d := \frac{x\sqrt{mk}\sqrt{2}\sqrt{gh}}{mg}$$

The result for d, v, and t after this sequence is the same as if the expression given by Eq. 2.33 had been supplied as the argument to the assign command, except for the manual simplifications that were done earlier.

Returning to the analysis of the solution, Eq. 2.33 shows that the horizontal distance travelled by the ball is directly proportional to the amount that the spring is compressed. Thus, in order to increase the distance travelled by 25% (to get from 1.6 m to 2.0 m), the professor should increase the amount of spring compression by 25%, to 1.25 cm.

Follow the one-equation-at-a-time prescription to find d as a function of x for the case in which the ball is projected upward at some angle θ with respect to the horizontal. Verify that your result agrees with the above in the $\theta = 0$ limit.

2.5   Momentum Conservation

The component of the total momentum of a system of particles is constant in any direction in which the corresponding component of the net external force is zero. Furthermore, even when the net external force is not zero, for some problems the time of interest (*e.g.*, the duration of a collision) is sufficiently short that external forces cannot change the total momentum significantly during that time. Like conservation of mechanical energy, momentum conservation can be used to get information about the dynamics of one or more objects without having to consider the details of the forces on the objects. This is particularly useful for problems involving collisions, as the following examples illustrate.

2.5.1   A Head-On Elastic Collision

We first examine the problem of a mass m, initially moving with speed v_0, which collides head-on with a mass M initially at rest. Let us assume that the collision is elastic, so that kinetic energy is conserved. If the velocities of m and M after the collision are v and V, respectively, conservation of momentum and kinetic energy lead to

$$mv_0 = mv + MV$$

and

$$\frac{1}{2}mv_0^2 = \frac{1}{2}mv^2 + \frac{1}{2}MV^2 \ .$$

These two equations are loaded into Maple in the usual manner, and solved for v and V with the solve command.

```
> restart;
> eq1 := m*v0 = m*v + M*V:
> eq2 := 1/2*m*v0^2 = 1/2*m*v^2 + 1/2*M*V^2:
> sol := solve({eq1,eq2}, {v,V});
```

$$sol := \{v = v0, \ V = 0\}, \ \left\{ v = \frac{(m-M)\,v0}{m+M}, \ V = 2\frac{m\,v0}{m+M} \right\}$$

The first set corresponds to a solution to the equations at a time prior to the collision. Thus, the solution set that we want is the second one. We assign the corresponding values to v and V so that we can conveniently analyze the solution.

```
> assign(sol[2]);
```

Several obvious limits to examine with Maple's limit command are listed below with the results for v and V:

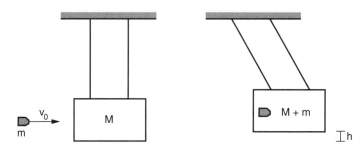

FIGURE 2.9. The ballistic pendulum, before and after being struck by an incoming bullet.

$$
\begin{aligned}
M \to \infty: \quad & v = -v_0, \quad V = 0 \\
m = M: \quad & v = 0, \quad V = v_0 \\
m \to \infty: \quad & v = v_0, \quad V = 2v_0 \, .
\end{aligned}
$$

In general, if the incoming mass is smaller than the target mass ($m < M$), it is reflected back in the opposite direction ($v < 0$). If $m = M$ it stops, and if $m > M$, it continues in the same direction, though at a lesser speed. In all cases after the collision the target mass moves in the original direction of the incoming mass with a speed in the range $0 < V < 2v_0$.

2.5.2   The Ballistic Pendulum

The ballistic pendulum, shown in Fig. 2.9, is a large block suspended in such a way that it is free to swing. It can be used to determine the speed of a bullet of known mass, using conservation of momentum and conservation of energy. The analysis proceeds as follows. The bullet comes in with speed v_0 and collides with the block in a totally inelastic collision. Immediately after the collision the block and embedded bullet move to the right with speed V, which can be determined using conservation of momentum, assuming the block does not move significantly during the time it takes the bullet to come to rest in it. (This insures that during the collision the net external force in the x-direction is approximately zero.) After the collision mechanical energy is conserved as the block and embedded bullet rise to a height h above the original height of the block, with the kinetic energy being changed to gravitational potential energy. Note that the initial kinetic energy of the bullet cannot be equated to the final potential energy of the block and embedded bullet because mechanical energy is not conserved during the inelastic collision.

To solve this problem we find the functional dependence of the height h upon the initial speed v_0. The momentum conservation equation for the collision is

$$
mv_0 = (m + M)V \, . \tag{2.34}
$$

After the collision energy conservation provides the equation

$$\frac{1}{2}(m + M)V^2 = (m + M)gh . \tag{2.35}$$

Using Maple to solve Eqs. 2.34 and 2.35 simultaneously for h and V yields

$$V = \frac{m}{m + M}v_0 \tag{2.36}$$

and

$$h = \frac{m^2 v_0^2}{2g(m + M)^2} . \tag{2.37}$$

This result for h is obtained after a call to the factor command. If we wish to use the measured height to compute the speed of the bullet, we need only solve Eq. 2.37 for v_0.

Verify that the expressions for h and V have the correct $m \to 0$, $M \to \infty$, and $v_0 \to 0$ limits.

It is interesting to compare the ratio of the mechanical energy of the block and embedded bullet after the collision to the initial kinetic energy of the bullet. Manipulating Eq. 2.37 provides this ratio as

$$\frac{(m + M)gh}{\frac{1}{2}mv_0^2} = \frac{m}{m + M} .$$

If $m \ll M$, as is typical, we see that a large amount of the initial kinetic energy is dissipated as heat.

Finally, what would happen if the bullet collided elastically with the block? Since $m < M$, the example in the Sec. 2.5.2 indicates that the bullet would bounce back, thus giving more momentum to the block than in the inelastic case. In fact, comparison with that example reveals that V is twice as large for an elastic collision as for a totally inelastic collision. Thus a bullet striking the block elastically would cause the block to rise to four times the height reached in the totally inelastic case.

2.5.3 A Collisional Party Trick

Suppose we drop a large ball of mass M and a small ball of mass $m \ll M$ simultaneously, with the small ball on top of the large ball, as in Fig. 2.10. The balls fall from an initial height H. If the collisions are approximately elastic, the small ball will rebound to a height nearly nine times its initial height! Let us analyze this problem, making the assumptions that the collisions are

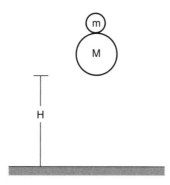

FIGURE 2.10. Two balls dropped simultaneously from an initial height H.

elastic and of such short duration that the gravitational force changes the total momentum negligibly during the collision time.

It is convenient to think of the motion as occurring in several stages. First, the two balls fall under the influence of gravity, changing their initial potential energy into kinetic energy. Letting v_0 be their speed just before the large ball hits the ground, energy conservation gives

$$(m + M)gH = \frac{1}{2}(m + M)v_0^2 . \tag{2.38}$$

In the second stage, the large ball collides elastically with the ground, causing it to reverse its velocity from v_0 to $-v_0$. It then collides elastically with the small ball, with the resulting velocities of V and v for the large and small balls, respectively. This collision is described by the equations

$$-Mv_0 + mv_0 = MV + mv \tag{2.39}$$

and

$$\frac{1}{2}Mv_0^2 + \frac{1}{2}mv_0^2 = \frac{1}{2}MV^2 + \frac{1}{2}mv^2 . \tag{2.40}$$

Finally, in the last stage, the small ball, which is travelling with a velocity v upward after its collision, rises to a height h as its kinetic energy is changed to potential energy:

$$\frac{1}{2}mv^2 = mgh . \tag{2.41}$$

To solve, we enter Eqs. 2.38 through 2.41 into Maple as eq1, eq2, eq3, and eq4.

```
> restart;
> eq1 := (m + M)*g*H = 1/2*(m + M)*v0^2:
> eq2 := - M*v0 + m*v0 = M*V + m*v:
> eq3 := 1/2*M*v0^2 + 1/2*m*v0^2 = 1/2*M*V^2 + 1/2*m*v^2:
```

> eq4 := 1/2*m*v^2 = m*g*h:

To avoid dealing with Maple RootOf expressions, we follow the step-by-step procedure of solving an equation for one unknown in terms of the others, and substitute the result into the remaining equations, choosing the appropriate root when necessary. We repeat the process until we have one remaining equation which we can solve for h. So, for example, we first solve Eq. 2.38 for v_0. There are two roots; assuming upward is the positive direction, we must choose the negative root for v_0.

> solve(eq1, v0);

$$\frac{1}{2}\frac{\sqrt{2}\sqrt{gH}(m+M)}{-\frac{1}{2}m-\frac{1}{2}M}, \quad -\frac{1}{2}\frac{\sqrt{2}\sqrt{gH}(m+M)}{-\frac{1}{2}m-\frac{1}{2}M}$$

> v0 := simplify("[1]);

$$v0 := -\sqrt{2}\sqrt{gH}$$

The assignment of the value to v0 automatically makes the substitution into the remaining three equations.

Now we solve the modified Eq. 2.39 for V.

> solve(eq2, V):
> V := simplify(");

$$V := -\frac{-M\sqrt{2}\sqrt{gH}+m\sqrt{2}\sqrt{gH}+mv}{M}$$

The resulting modification of Eq. 2.40 is solved for v with

> solve(eq3, v);

Maple output

The output from this command, which we have simply denoted as "Maple output," is a sequence of two roots, only one of which is physically acceptable. To discover which one, we apply

> simplify(["]);

$$\left[-\sqrt{2}\sqrt{gH}, \quad -\frac{\sqrt{2}\sqrt{gH}(-3M+m)}{m+M}\right]$$

The brackets convert the two-element sequence returned by solve into a list. This is necessary because simplify expects a single argument.

The first root corresponds to the pre-collision solution. Thus we assign the second root to v.

> v := "[2]:

Finally, we solve Eq. 2.41 for the height h. The value for v is automatically substituted by Maple. The Maple statements

> solve(eq4, h);

Maple output

> simplify(");

$$\frac{H\left(9M^2 - 6Mm + m^2\right)}{(m+M)^2}$$

> h := factor(");

complete the solution, yielding the following result for h.

$$h = \frac{H(m-3M)^2}{(m+M)^2}$$

With foresight we could have combined these three statements into a single one, of course.

From the final expression for h, we obtain the surprising result mentioned in the original problem statement by examining the $m \to 0$ limit:

> limit(h, m=0);

$$9H$$

In the same limit, $V \to \sqrt{2gH}$, which means that the large mass returns almost to its original height.

The experiment is tricky to perform because it is extremely difficult to insure that the small ball hits the large one along a line passing through the center of the large one. Dropping them in a tube helps insure that the rebound of the small ball is directed vertically. As for the balls themselves, a ping pong ball will do for m, whereas a racketball or tennis ball serves for M. The inherent inelasticity of these balls, however, prevents the small one from achieving the full 9-fold increase in height. Nevertheless, the effect is startling.

2.6   Problems

Note: In the problems below, when asked to "verify" a limit you are expected to not only find the limit, but to explain physically why the limiting result is correct.

1. Consider a projectile launched at an angle θ above the horizon. Neglecting air resistance,

    (a) Obtain the expression for the range of a projectile as a function of θ;

    (b) Show that the maximum range occurs when $\theta = \frac{\pi}{4}$;

    (c) Show that the range of the projectile when launched at an angle $\theta = \frac{\pi}{4} + \alpha$ is the same as it would be if it were launched at $\theta = \frac{\pi}{4} - \alpha$.

2. Imagine yourself at a third floor window of a burning building, a height h above the ground. You have a rope which is long enough to reach the ground, but it will support a maximum tension (T_m) which is less than your weight (W).

    (a) Find the minimum acceleration that you can slide down in order to keep from breaking the rope.

    (b) With what speed will you hit the ground if you accelerate at the minimum rate? How long will it take you to reach the ground?

    (c) Verify the $T_m = 0$ and $T_m = W$ limits for the acceleration, speed, and time.

3. A box of mass m_1 is being pulled up an incline of angle ϕ by a force F. It is connected to a second mass m_2 which is on a horizontal table, but just about to start up the incline. This second mass is in turn connected to a third mass m_3, also on the horizontal table. (See Fig. 2.11)

    (a) Write down Newton's 2nd law equations of motion for each of the three masses.

    (b) Find the common acceleration and the tensions in the ropes between the masses.

    (c) Verify the $\phi \to 0$ limit for the acceleration.

    (d) Explain the $\phi \to \frac{\pi}{2}$ limit for the acceleration.

    (e) Verify the $m_2 = m_3 \to 0$ limits for the acceleration and the tensions in the ropes.

4. A conical pendulum consists of a mass M at the end of a string of length L. The other end of the string is fixed. Assume the mass is moving in circular motion in a horizontal plane at an angular speed of ω.

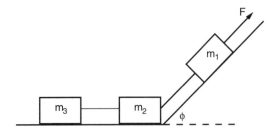

FIGURE 2.11. Problem 3.

(a) Manually solve for the angle between the taught string and the vertical.

(b) Repeat the calculation for the angle with Maple. If your manual calculation did not yield all of the solutions found by Maple, examine it to find what implicit assumption(s) you made that caused you to overlook one or more of the solutions.

(c) Discuss the physics of each of the solutions. Among other things you should indicate whether the solution is stable or unstable with regard to small displacements from the corresponding angle. Pay particular attention to the solution for which the angle depends upon ω; notice that there is a minimum value for ω in order for the solution to be real (and hence physically acceptable).

5. Starting from rest, a roller coaster car rolls without friction down the track, and then goes around a vertical, circular loop of radius R. The starting point is a vertical height H above the bottom of the loop.

(a) Find the minimum value of H required for the car to reach the top of the loop without falling away from the track.

(b) In terms of H, find the force that the track exerts on the car when it is upside down at the top of the loop, assuming $H > H_{min}$.

(c) At what height would the car fall away from the track loop if $2R < H < H_{min}$?

6. A wedge of mass M and angle α rests on a horizontal table. A block of mass m is placed on the wedge so that its lower edge is at a distance L from the table, as measured along the inclined surface. All surfaces are frictionless, and both the block and the wedge start from rest when the block is released.

(a) Energy and the horizontal component of momentum are conserved for this system. Why?

(b) Find the velocities of the wedge and the block (relative to the table) just as the corner of the block hits the table.

(c) Verify the $\alpha \to 0$ and $\alpha \to \frac{\pi}{2}$ limits for the results of the previous part.

7. An electron moving with speed v_0 collides head-on with an atom which is initially at rest. After the collision the atom is left in an excited state, internally storing an energy of amount E.

(a) Find the mathematically possible final velocities of the electron and atom. Note that there will be two solution sets.

(b) Using the fact that the velocities must be real, find the minimum value of v_0 that will allow the atom to absorb energy E. At first thought you might expect the minimum v_0 to be equal to $\sqrt{2E/m}$. Explain physically why, in fact, it is larger.

3 Newtonian Dynamics of Particles

3.1 Kinematics in Other Coordinate Systems

Although in principle any problem involving classical mechanics of a particle can be expressed in terms of rectangular coordinates, for some problems other kinds of coordinate systems may result in simpler formulations. In this section we look at three other common coordinate systems that are quite useful, particularly for problems with axial or spherical symmetry. The discussion is confined to the kinematic equations for position, velocity, and acceleration.

3.1.1 Polar Coordinates

We consider first the x-y plane and the polar coordinate system which expresses the location of a point in terms of distance to the point, ρ, and angular measure ϕ from the x-axis. Figure 3.1 illustrates this system and its relationship to a rectangular coordinate system, in which a point is specified by giving its (x, y) coordinates. The relationship between rectangular and polar coordinates is given by

$$x = \rho \cos \phi ,$$

$$y = \rho \sin \phi .$$

These can be inverted to obtain

$$\rho = \sqrt{x^2 + y^2} ,$$

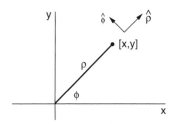

FIGURE 3.1. Illustration of polar coordinates.

$$\phi = \arctan\left(\frac{y}{x}\right) .$$

It is convenient to define two new unit vectors associated with the polar coordinates ρ and ϕ. The direction of the $\hat{\rho}$ unit vector is the same as that of the position vector $\mathbf{r}$; that is, radially outward from the origin. The $\hat{\phi}$ unit vector is perpendicular to $\hat{\rho}$, in the direction corresponding to an increase in the angle ϕ (counterclockwise). This is sometimes referred to as the tangential direction, since it is tangent to a circle of radius ρ centered at the origin. Note that the directions of these two unit vectors change as the angle changes. Their directions are shown in Fig. 3.1 for the particular point (x,y).

In terms of polar coordinates, the position vector for the given point is

$$\mathbf{r} = \rho\,\hat{\rho} . \tag{3.1}$$

Even though it may appear that the position vector depends on the single coordinate ρ, remember that the unit vector depends on ϕ. By comparing this polar coordinate expression for the position vector with the rectangular coordinate expression,

$$\mathbf{r} = x\,\hat{\mathbf{x}} + y\,\hat{\mathbf{y}} ,$$

we find the following relationship between $\hat{\rho}$ and the rectangular unit vectors $\hat{\mathbf{x}}$ and $\hat{\mathbf{y}}$.

$$\hat{\rho} = \cos\phi\,\hat{\mathbf{x}} + \sin\phi\,\hat{\mathbf{y}} \tag{3.2}$$

This explicitly shows the ϕ dependence of the unit vector. The unit vector $\hat{\phi}$, which is perpendicular to $\hat{\rho}$, represents the change in the direction of $\mathbf{r}$ when ϕ changes, holding ρ constant. A derivation may be found in standard classical mechanics texts; see, e.g., [15]. The result is

$$\hat{\phi} = \frac{d\hat{\rho}}{d\phi} = -\sin\phi\,\hat{\mathbf{x}} + \cos\phi\,\hat{\mathbf{y}} . \tag{3.3}$$

Verify that $\hat{\phi}$ is perpendicular to $\hat{\rho}$.

With expressions for the unit vectors now established, we can find the velocity and acceleration of a particle in a plane in terms of polar coordinates. Starting with Eq. 3.1 for the position, we differentiate with respect to time to find the velocity:

$$\mathbf{v} = \frac{d\mathbf{r}}{dt}$$

$$= \frac{d\rho}{dt}\hat{\rho} + \rho\frac{d\hat{\rho}}{dt} .$$

Applying the chain rule,

$$\frac{d\hat{\rho}}{dt} = \frac{d\hat{\rho}}{d\phi}\frac{d\phi}{dt} = \dot{\phi}\hat{\phi} , \tag{3.4}$$

we find that the velocity has both $\hat{\rho}$ and $\hat{\phi}$ components:

$$\mathbf{v} = \dot{\rho}\hat{\rho} + \rho\dot{\phi}\hat{\phi} . \tag{3.5}$$

The acceleration is found by differentiating this expression for the velocity. There are multiple terms, since everything on the right side of the equation can change with time:

$$\mathbf{a} = \ddot{\rho}\hat{\rho} + \dot{\rho}\frac{d\hat{\rho}}{dt} + \dot{\rho}\dot{\phi}\hat{\phi} + \rho\ddot{\phi}\hat{\phi} + \rho\dot{\phi}\frac{d\hat{\phi}}{dt} . \tag{3.6}$$

The time derivative of $\hat{\phi}$ is evaluated as it was for $\hat{\rho}$:

$$\frac{d\hat{\phi}}{dt} = \frac{d\hat{\phi}}{d\phi}\frac{d\phi}{dt} = -\dot{\phi}\hat{\rho} . \tag{3.7}$$

Substituting Eqs. 3.4 and 3.7 into Eq. 3.6, and collecting $\hat{\rho}$ and $\hat{\phi}$ terms reveals that the two-dimensional acceleration expressed in terms of polar coordinates is given by

$$\mathbf{a} = (\ddot{\rho} - \rho\dot{\phi}^2)\,\hat{\rho} + (\rho\ddot{\phi} + 2\dot{\rho}\dot{\phi})\,\hat{\phi} . \tag{3.8}$$

To investigate how Eqs. 3.5 and 3.8 apply to the case of circular motion at constant speed, we assume that the origin of the polar coordinate system is at the center of the circle. If the motion is indeed circular, ρ is constant, so that $\dot{\rho} = 0$. Furthermore, if the speed is constant, Eq. 3.5 indicates that the angular speed $\dot{\phi} = \omega$ is constant. Thus, the velocity and acceleration of a particle in uniform circular motion are given by

$$\mathbf{v} = \rho\omega\,\hat{\phi}$$

and

$$\mathbf{a} = -\rho\omega^2\,\hat{\rho} .$$

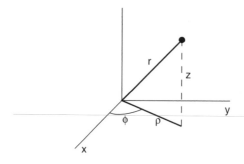

FIGURE 3.2. Illustration of cylindrical coordinates.

As expected, the velocity is in the tangential direction, with a speed of $\rho\omega$. The acceleration is toward the center of the circle $(-\hat{\rho})$ with a magnitude given by

$$a = \rho\omega^2 = \frac{v^2}{\rho} .$$

This is the well-known centripetal acceleration.

3.1.2   Cylindrical Coordinates

In three dimensions, an obvious extension of polar coordinates is the addition of a z-coordinate whose axis is perpendicular to the x-y plane in which polar coordinates were defined. The position vector then has a z-component in addition to the ρ-component of the previous section. Thus,

$$\mathbf{r} = \rho\,\hat{\rho} + z\,\hat{\mathbf{z}} , \tag{3.9}$$

where ρ and $\hat{\rho}$ are defined as in the preceding section. See Fig. 3.2.

Since $\hat{\mathbf{z}}$ is constant and perpendicular to both $\hat{\rho}$ and $\hat{\phi}$, the cylindrical coordinate expressions for velocity and acceleration are the same as those for polar coordinates, except for the addition of z-components:

$$\mathbf{v} = \dot{\rho}\,\hat{\rho} + \rho\dot{\phi}\,\hat{\phi} + \dot{z}\,\hat{\mathbf{z}} ,$$

and

$$\mathbf{a} = (\ddot{\rho} - \rho\dot{\phi}^2)\,\hat{\rho} + (\rho\ddot{\phi} + 2\dot{\rho}\dot{\phi})\,\hat{\phi} + \ddot{z}\,\hat{\mathbf{z}} .$$

3.1.3   Spherical Coordinates

In spherical coordinates, the position vector of a point is written as the distance (r) from the origin to the point, times a unit vector directed from the origin to the point; that is,

$$\mathbf{r} = r\hat{\mathbf{r}} .$$

The direction of the vector is specified by two angles: the angle that $\mathbf{r}$ makes with the z-axis (θ) and the angle that the projection of $\mathbf{r}$ onto the x-y plane makes with the x-axis (ϕ). The angle ϕ is the same as that in cylindrical and polar coordinates, and the cylindrical and polar coordinate ρ is equal to the projection of $\mathbf{r}$ onto the x-y plane. Associated with θ there is a unit vector $\hat{\theta}$ tangent to a sphere of radius r, in the direction of increasing θ. The unit vector $\hat{\phi}$ is the same as it is in cylindrical and polar coordinates, tangent to a circle of radius ρ and pointing in the direction of increasing ϕ. Thus, ϕ represents a rotation around the z-axis.

The coordinates of the rectangular system can be expressed in terms of r, θ, and ϕ with the three equations

$$x = r \sin\theta \cos\phi \, ,$$

$$y = r \sin\theta \sin\phi \, ,$$

$$z = r \cos\theta \, .$$

These equations can be used to obtain the spherical coordinate unit vectors $\hat{\mathbf{r}}$, $\hat{\theta}$, and $\hat{\phi}$. The velocity and acceleration can be obtained from the position vector by taking respectively the first and second derivatives with respect to time. Spherical kinematic expressions in spherical coordinates are derived in standard classical mechanics texts, such as[15]. They are not used in this text, so we do not give the results here. However, they are quite useful for advanced topics, such as describing the motion of masses relative to a coordinate system fixed on the earth's surface when the rotation of the earth is important.

3.2   Explicitly Time-Dependent Forces

The previous chapter looked at several examples of applications of Newton's 2nd law in which the applied forces were constant. When the net force on a particle is constant, we can solve for the acceleration, and integrate it to find the velocity and position as functions of time, as described in Section 2.1. If the net force is given as an explicit function of time, we can still use Newton's 2nd law to solve for the acceleration, and again integrate to find the velocity and position as functions of time. The integrations are more difficult because of the time-dependence of the acceleration. In fact, for some cases it may not be possible to obtain explicit expressions for the velocity and/or position; we may have to do the integrations numerically. Nevertheless, the procedure is straightforward: Solve for the acceleration. Integrate it to find the velocity. Integrate the velocity to find the position.

An equivalent way to deal with motion of a particle under explicitly time-dependent forces is to integrate both sides of Eq. 2.6 over time to obtain

$$\int_{t_0}^{t} dt' \, \mathbf{F}(t') = \mathbf{p}(t) - \mathbf{p}(t_0) \, .$$

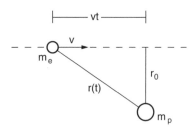

FIGURE 3.3. Impulsive collision between an incoming electron and a proton at rest.

The integral of the force over time is known as the impulse, and this result is aptly named the *impulse-momentum theorem*.

As an example of its use, we consider an electron passing a proton which is at rest. Let us assume that the speed of the electron is so large that its velocity is changed negligibly by the interaction with the proton; that is, the electron path is nearly a straight line, and its speed is approximately constant. The conditions for the validity of this approximation are discussed later. We assume also that because the proton is so much more massive than the electron it does not move significantly until the electron has passed. A schematic of the problem is shown in Fig. 3.3. The electron comes in with speed v, passes within a distance r_0 of the proton, and then continues on. The time when the electron is closest to the proton is taken to be $t = 0$.

Under the assumptions made, the distance from the electron to the proton is approximately

$$r(t) = \sqrt{r_0^2 + v^2 t^2} \; .$$

We can use Coulomb's law to obtain the component of the force that the electron exerts on the proton in the direction perpendicular to the path of the electron. If the x-axis is along this perpendicular direction, then (in SI units)

$$F_x(t) = \frac{e^2 r_0}{4\pi\epsilon_0 \left[r_0^2 + v^2 t^2\right]^{3/2}} \; . \tag{3.10}$$

The x-component of the impulse delivered by the electron to the proton is found by integrating this expression over time. The result is the change in the x-component of the proton's momentum. Thus the x-component of its velocity long after the electron has passed is approximately given by

$$V_x = \frac{1}{m_p} \int_{-\infty}^{\infty} dt \, F_x(t) \; . \tag{3.11}$$

The following sequence solves the problem, with M representing the mass of the proton, m_p.

```
> k := e^2/(4*Pi*epsilon):
```

```
> Fx := k*r0/( r0^2 + (v*t)^2 )^(3/2):
> Impulse := int(Fx, t=-infinity..infinity):
> Vx := Impulse/M;
```

Since $v>0$, the result is

$$V_x = \frac{e^2}{2\pi\epsilon_0 r_0 m_p v} \, . \tag{3.12}$$

Show that the y-component of the impulse and velocity are zero under the approximations made.

If the external forces on the electron-proton system are neglected, the x-component of the electron's momentum is the negative of the proton's after the collision, $m_e v_x = -m_p V_x$. Thus, the x-component of the electron's velocity is approximately

$$v_x = \frac{-e^2}{2\pi\epsilon_0 r_0 m_e v} \, .$$

We can use this result to check the assumption that the electron continues to move approximately in a straight line. We require that

$$\frac{|v_x|}{v} = \frac{e^2}{2\pi\epsilon_0 r_0 m_e v^2} \ll 1 \, . \tag{3.13}$$

Physically, Eq. 3.13 states that the electron will not deviate significantly from its straight-line path if its potential energy at the point of closest approach is much smaller than its kinetic energy.

To check the approximation that the proton does not move significantly during the time of interaction with the electron, we let $2T$ be a time interval over which most of the impulse is derived. For $|t| > T = 10r_0/v$, for example, the force is less than 10^{-3} times its maximum value. Moreover, Maple enables us to show that the interval from $-T$ to T contributes more than 99% of the impulse integral.

```
> T := 10*r0/v:
> assume(v>0):   assume(r0>0):
> evalf(simplify(int(Fx, t=-T..T)/Impulse));
```

.9950371901

A conservative test of the approximation is to verify that $V_x(2T) \ll r_0$.

We can employ Maple to show

$$\frac{2V_x T}{r_0} = \frac{10e^2}{\pi\epsilon_0 r_0 m_p v^2}$$

$$= \left(\frac{20m_e}{m_p}\right) \frac{e^2}{2\pi\epsilon_0 r_0 m_e v^2} . \tag{3.14}$$

Since the mass of the proton is about 1836 times that of the electron, the factor in parentheses is substantially less than 1. Moreover, in order for the electron path to be approximately straight, Eq. 3.13 indicates that the second factor must be much less than 1. Hence, the approximation that the proton does not move significantly during the interaction is quite good so long as the electron is moving fast enough that its path is nearly straight.

3.3   Position- or Velocity-Dependent Forces

For most problems of physics the forces are not known explicitly as functions of time. Rather, the forces are typically given as functions of position and/or velocity, and so depend on time indirectly. For these types of problems Newton's second law results in one or more differential equations which must be solved to find $r(t)$ or $v(t)$. In this section we study some simple problems of this type, using Maple to solve the equations and to help analyze the physics of the problem.

3.3.1   Projectile Motion with Air Resistance

As the first example, let us consider the motion of a projectile when air resistance is not negligible. Empirically, the drag on a projectile caused by its motion through the air varies directly with the speed of the object, and in a direction opposite to the velocity. The drag force is a complex function of velocity depending on the shape of the object, among other things. However, a reasonable approximation to the actual force due to air resistance, or the resistive force on any object moving through a viscous medium, is one which is directly proportional to velocity, $F_{drag} = -bv$. The parameter b is the drag coefficient, which is a measure of the resistance of the medium to the passage of the given object. We apply this model to calculate the position and velocity of a projectile as functions of time, and to examine the effect of drag upon the trajectory.

We assume that x and y denote the horizontal and vertical directions, respectively. Since neither of the two forces that act on the object depend explicitly on the position, Newton's second law can be used to obtain an equation of motion which is a first order differential equation for $v(t)$, rather than the more common second order equation for $\mathbf{r}(t)$:

$$-mg\,\hat{\mathbf{y}} - b\mathbf{v} = m\frac{d\mathbf{v}}{dt}\ .$$

With some rearrangement, this vector equation can be broken into the x and y equations

$$m\frac{dv_x(t)}{dt} + bv_x(t) = 0 \tag{3.15}$$

and

$$m\frac{dv_y(t)}{dt} + bv_y(t) + mg = 0\ . \tag{3.16}$$

The initial values of velocity are taken to be $v_x(0) = v_{0x}$ and $v_y(0) = v_{0y}$.

To solve Eqs. 3.15 and 3.16 with Maple, we first use diff to enter the equations,

```
> eqx := m*diff(vx(t), t) + b*vx(t) = 0:
> eqy := m*diff(vy(t), t) + b*vy(t) + m*g = 0:
```

and then solve them with dsolve. Although the equations are not coupled, and thus can be solved independently, we solve them together to illustrate the method for solving coupled equations.

The first argument of dsolve is the set of equations to be solved. This set may include equations which specify the initial conditions. The second argument is the set of dependent variables. The solutions are found with the Maple statement

```
> sol := dsolve({eqx, eqy, vx(0)=v0x, vy(0)=v0y}, {vx(t), vy(t)});
```

They are

$$v_x(t) = v_{0x}e^{-bt/m} \tag{3.17}$$

and

$$v_y(t) = -\frac{mg}{b} + \left(v_{0y} + \frac{mg}{b}\right)e^{-bt/m}\ . \tag{3.18}$$

We check these solutions with the same Maple command that we used to check the solutions to a set of algebraic equations.

```
> simplify(subs(sol,{eqx,eqy}));
```

$$\{0 = 0\}$$

Like solve for algebraic equations, the dsolve command does not assign values to the variables x(t) and y(t). This is done with assign as before. The variables can then be checked to insure that they satisfy the initial conditions.

```
> assign(sol);
> limit(vx(t), t=0);
```
$$v0x$$

```
> limit(vy(t), t=0);
```
$$v0y$$

As always, it is important to verify limiting cases for which the answers are known or can be inferred. One obvious case to check is that of no air resistance, $b = 0$. This case must, of course, agree with the standard result for projectile motion.

```
> limit(vx(t), b=0);
```
$$v0x$$

```
> limit(vy(t), b=0);
```
$$-gt + v0y$$

By examining a small time expansion of Eqs. 3.17 and 3.18 we see that the velocity components start at their initial values and begin to decrease linearly with time — vx(t) because of the air resistance, and vy(t) due to the joint effect of gravitational attraction and air resistance, assuming the object is initially projected upward. This short-time behavior is obtained with Maple by using the series command.

```
> series(vx(t), t, 3);
> series(vy(t), t, 3);
```

which yield

$$v_{0x} - \frac{v_{0x}b}{m}t + \frac{1}{2}\frac{v_{0x}b^2}{m^2}t^2 + O(t^3)$$

and

$$v_{0y} - \frac{mg + v_{0y}b}{m}t + \frac{1}{2}\frac{(mg + v_{0y}b)b}{m^2}t^2 + O(t^3).$$

At longer times, the deviation from the no-air-resistance case becomes more pronounced because of the exponential factors in Eqs. 3.17 and 3.18. When $t \gg m/b$, the exponentials are negligible, so that

$$v_x(t) \to 0$$

and

$$v_y(t) \to -\frac{mg}{b} \tag{3.19}$$

in the long-time limit, assuming that the object has not yet struck the ground. Maple can verify these limits if we tell it that m/b is positive using assume. That is,

```
> assume(m>0):    assume(b>0):
> limit(vx(t), t=infinity);
```

$$0$$

```
> limit(vy(t), t=infinity);
```

$$-\frac{\tilde{m}\,g}{\tilde{b}}$$

The limit in 3.19 is the well-known "terminal velocity." Physically it arises because as a falling object picks up speed due to the constant gravitational force, the resistive force grows in magnitude. Eventually the two forces cancel, resulting in zero vertical acceleration, and thus constant downward component of velocity. This result is easily obtained by solving Eq. 3.16 when $dv_y/dt = 0$.

Maple treats vx(t) and vy(t) as variable names; it does not recognize them to be functions of t. So, for example, vx(0) will not yield the initial value for vx(t). However, in order to graphically compare the velocity and position as functions of time for different drag coefficients, it is convenient to create functions of b and t out of the expressions which are the values of the Maple variables vx(t) and vy(t). We call on unapply for this task. We first remove the assumptions on b and m and then enter the unapply command.

```
> b := 'b':   m := 'm':
> VX := unapply(vx(t), b, t);
```

$$VX := (b, t) \rightarrow v0x\, e^{-\left(\frac{bt}{m}\right)}$$

As seen in Chap. 1, the notation on the right side of the assignment operator in the output is what Maple uses to indicate a function, in this case a function of the two variables b and t. In the other expressions in this Maple session b and t are global variables. In this expression they are local variables, representing the formal parameters of the VX function.

In a similar fashion we define the function which represents the y-component of the velocity.

```
> VY := unapply(vy(t), b, t):
```

Capital letters have been used for VX and VY so as not to confuse them with the variables vx(t) and vy(t).

Assuming the origin of the coordinate system is at the initial position of the object, the x- and y-components of position may be defined by the Maple assignments

```
> X := unapply(int(vx(t), t=0..s), b, s):
> Y := unapply(int(vy(t), t=0..s), b, s):
```

Note that we may not use the integration variable in the limits of an integration; this accounts for our introduction of s as one of the arguments for the X and Y functions.

As in many examples, plots of the velocity and position of the projectile are helpful in visualizing the motion and in understanding the physics of this problem. In order to make such plots we need to assign values to the parameters of the system. We first examine the special case in which the object is dropped vertically from rest. Depending on the value of b, this case can apply approximately to a rock dropped from the edge of a cliff or to a marble falling through a very viscous liquid such as honey. We make the following assignments (in SI units):

> m := 1:   g := 9.8:   v0x := 0:   v0y := 0:

Maple automatically incorporates these values into the position and velocity functions, allowing us to plot the results with little effort. With these initial conditions the motion is strictly one-dimensional because $v_x(t)$ and $x(t)$ both remain zero for all times.

We consider the y-component of the velocity as a function of time, for three different values of drag: $b = 0$, 1, and 10 kg/s. The Maple command

> plot({limit(VY(b,t),b=0), VY(1,t), VY(10,t)}, t=0..1);

plots these velocities over the time interval from 0 to 1 s. We require the $b = 0$ limit in the command because the definition for VY(b,t) leads to a division by 0 if we simply take VY(0,t).

The results are shown in Fig. 3.4. The $b = 0$ curve shows the expected linear dependence of the velocity upon time. Its slope can be estimated from the plot, and is in agreement with the -9.8 m/s^2 expected for free-fall. For the intermediate value of the drag coefficient ($b = 1$ kg/s), the speed deviates from linearity, as indicated by the small-time expansion earlier. At larger values of t the curve begins to vary substantially from the zero drag curve. Were we to plot it for still longer times it should eventually approach the terminal velocity given earlier. The $b = 10$ kg/s curve, which corresponds to a very viscous fluid, clearly shows the onset of terminal velocity. Its value can be estimated from the plot to be approximately -1 m/s, in agreement with the -0.98 m/s predicted by Eq. 3.19.

Plotting several curves on a single graph, as in Fig. 3.4, is a good way of getting a sense of the dependence of a quantity upon multiple parameters. An alternate method of investigating the b-dependence along with the t-dependence is to use the animate command, which is in the plots package. Although animate is commonly used to show the evolution of a system in time, it can also be used to step through a series of curves by varying a specified parameter. In the present context, the command

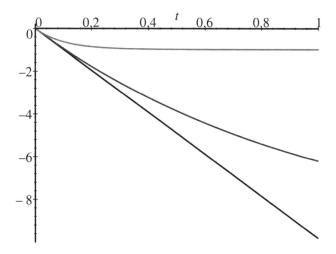

FIGURE 3.4. Velocity as a function of time for an object falling through a viscous fluid.

> plots[animate](VY(b,t), t=0..1, b=0..10);

will produce a series of $v_y(t)$ vs. t frames. The frames, each representing a different b-value, can be displayed one at a time, or in an automated sequence. By watching the evolution of the curves with b we can visualize the dependence of particle's velocity as a function of the drag coefficient.

Another technique for viewing the dependence of the particle's velocity upon b and t is to do a 3-dimensional plot. Maple's plot3d command creates such a plot. For example, Fig. 3.5 was obtained from the command

> plot3d(VY(b,t), t=0..1, b=0..10, orientation=[-60,75], axes =NORMAL);

The fourth argument, orientation=..., specifies the viewing angles. (See ?plot3d and ?options[plot3d].) However, since the orientation of the plot is easily adjusted interactively, it is not necessary to include this optional argument.

Although it may take some study to become comfortable with the three-dimensional plot, it provides a more complete qualitative picture of the dependence of v_y upon b and t. However, it is very difficult to extract quantitative information from it. For that a plot similar to Fig. 3.4 is more appropriate for any particular choice of b.

We note that in the solutions for $v_x(t)$ and $v_y(t)$, Eqs. 3.17 and 3.18, the parameter b appears everywhere in the form of the ratio, b/m. The same is true for $x(t)$ and $y(t)$. A consequence of this is that we need not change the value of m and plot anew to understand the effect of drag on a more massive or less massive object. Instead, we can retain $m = 1$ and plot for b values scaled to have the desired b/m ratio. Thus, for a given value of b, we

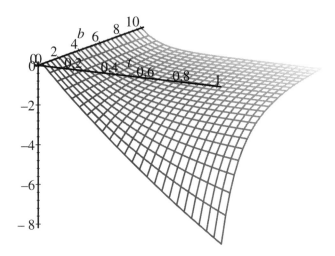

FIGURE 3.5. Velocity as a function of time and drag coefficient for an object falling through a viscous fluid.

need only look at VY(b/2,t) to see the effect of doubling m. The ratio m/b, which has units of time, can be thought of as a characteristic time for this problem — a measure of how rapidly $v_x(t)$ goes to zero or $v_y(t)$ approaches its terminal velocity. A full discussion of scaling and characteristic units is presented in Chapter 4.

Examine the truth of the scaling argument outlined in the preceding paragraph. For example, compare an $m = 1$ kg, $b = 2$ kg/s plot with an $m = 2$ kg, $b = 4$ kg/s plot. Try other appropriate combinations.

In order to see the effect of drag upon the two-dimensional motion of a projectile, we change the values of v_{0x} and v_{0y} to 10 m/s.

> v0x := 10:   v0y := 10:

The new parameter values are automatically incorporated in the evaluation of the VX(b,t), VY(b,t), X(b,t), and Y(b,t) functions since we made the earlier assignments for v0x and v0y after the functions were defined. After making these changes in the initial velocity, we call upon Maple's parametric plotting capability to plot the trajectory of the object, y(t) *vs.* x(t), for several values of drag. The following Maple statement does this for the three values $b = 0$, 0.1, and 1 kg/s, over a time interval from 0 to 2 s.

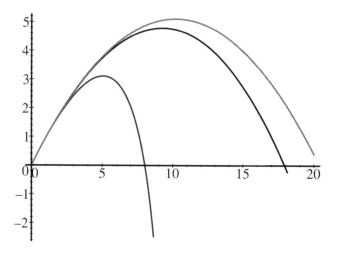

FIGURE 3.6. Trajectories of a projectile moving in a uniform gravitational field with a drag force proportional to velocity.

```
> plot({[limit(X(b,t),b=0), limit(Y(b,t),b=0), t=0..2],
>          [X(.1,t), Y(.1,t), t=0..2], [X(1,t), Y(1,t), t=0..2]});
```

The results are shown in Fig. 3.6. All curves have the same initial $45°$ slope because of the initial conditions on the velocity components. The $b = 0$ kg/s curve is parabolic, as appropriate for projectile motion with no air resistance. With a small, but non-zero value for b (= 0.1 kg/s), the trajectory deviates from parabolicity at longer times, becoming more nearly vertical as $v_x(t) \to 0$ and $v_y(t) \to -mg/b$. This effect is very pronounced in the $b = 1$ kg/s curve of the figure. As expected, air resistance affects both the range and maximum altitude of the projectile. Furthermore, if we interpret the $y(t) = 0$ axis as ground level, the time-of-flight is also seen to be reduced by drag.

Similar results can be found with different initial values. It is easy to change initial values or other parameters in Maple, and replot results. This interactivity is one of best reasons for using Maple to analyze physics.

The predicted behavior can be readily observed in a qualitative fashion by comparing the trajectories of a dense object like a baseball with a much less dense object such as a styrofoam ball. Although the mass of the styrofoam ball is much smaller than that of the baseball, the scaling property mentioned earlier suggests that its trajectory should have the same shape as that for a baseball moving through a fluid with a large b value. If you throw the styrofoam ball hard into the air, you should be able to

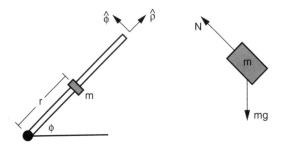

FIGURE 3.7. Schematic of and force diagram for a bead sliding on a rotating rod.

see it rapidly slow down, and drop almost vertically at the end
of its path.

3.3.2   Bead Sliding on a Rotating Rod

Figure 3.7 represents a rod, pivoted at one end, and a bead of mass m which
is allowed to slide along the rod. We assume that the rod moves in a vertical
plane with a constant angular acceleration α, and that it starts at rest from
a horizontal position. The angular position is thus given by $\phi = \frac{1}{2}\alpha t^2$. The
length of the rod is L, and initially the bead is at rest a distance ρ_0 from the
pivot point of the rod. We investigate the radial position of the bead as a
function of time.

The equation of motion for the bead is conveniently found by writing
Newton's second law for the radial direction, using the expression for the
radial acceleration in polar coordinates obtained earlier in this chapter. The
$\hat{\rho}$ equation is

$$F_\rho = m(\ddot{\rho} - \rho\dot{\phi}^2) . \tag{3.20}$$

Since the bead moves freely in the radial direction, $F_\rho = 0$, and the Maple
solution proceeds as follows.

```
> phi := 1/2*alpha*t^2;
> eq := 0 = diff(rho(t), t, t) - rho(t)*diff(phi,t)^2:
> sol := dsolve({eq, rho(0)=rho0, D(rho)(0)=0}, rho(t));
```

$$\rho(t) = \tfrac{1}{2}\rho0\,\alpha^{1/4}\,\Gamma\left(\tfrac{3}{4}\right)\sqrt{2}\sqrt{t}\,BesselI\left(\tfrac{1}{4}, \tfrac{1}{2}\alpha t^2\right)$$

$$+\tfrac{1}{\pi}\rho0\,\alpha^{1/4}\,\Gamma\left(\tfrac{3}{4}\right)\sqrt{t}\,BesselK\left(\tfrac{1}{4}, \tfrac{1}{2}\alpha t^2\right)$$

```
> simplify(subs(sol,eq));
```
$$0 = 0$$

```
> assign(sol);
```

The solution for $\rho(t)$ is not easily visualized in its complete form because of the Bessel functions. To help understand it better, let us look first at the small-time behavior.

> series(rho(t), t);

$$\rho0 + \frac{1}{12}\rho0\,\alpha^2\,t^4 + O\left(t^{13/2}\right)$$

Clearly the bead starts very slowly to move outward along the rod. However, our experience suggests that this outward motion will increase rapidly. We can verify this by looking at the asymptotic behavior of the solution for $\rho(t)$. The Maple asympt command, which is called in a very similar way to series, provides the first few terms of the asymptotic series. We use the optional third argument to restrict the result to the lowest order term.

> asympt(rho(t), t, 3);

$$\left(\frac{1}{2}\rho0\,\alpha^{1/4}\,\Gamma\left(\tfrac{3}{4}\right)\sqrt{2}\sqrt{\frac{1}{\pi\alpha}}e^{\left(\frac{1}{2}\alpha t^2\right)} + \frac{\rho0\,\alpha^{1/4}\,\Gamma\left(\tfrac{3}{4}\right)\sqrt{\frac{\pi}{\alpha}}e^{\left(-\frac{1}{2}\alpha t^2\right)}}{\pi}\right)\sqrt{\frac{1}{t}}$$
$$+O\left(\left(\tfrac{1}{t}\right)^{5/2}\right)$$

It is clear that due to the $\exp(\frac{1}{2}\alpha t^2)$ factor in the first term $\rho(t)$ increases extremely rapidly for larger times.

At what time does the bead leave the end of the rod? The answer can be found by solving for the value of t at which $\rho(t) = L$. This equation cannot be solved analytically, but the Maple fsolve command can give us numerical values if we specify α, ρ_0, and L. We can help it out by using a plot of $\rho(t)$ and L to narrow the domain in which the solution is to be found. For example,

> alpha := 1:   rho0 := L/2:   L := 1:
> plot({rho(t), L}, t=0..2);

The result, shown in Fig. 3.8, indicates that for these conditions the time t_f at which the bead leaves the end of the rod falls between 1.5 and 2.0 s if the values for α and L are in SI units. fsolve gives us a more accurate value for t_f, which can then be used to find the corresponding angular position of the rod. Note that, apparently due to a bug, Maple does not fully evaluate assigned solutions to differential equations, so we must use eval(rho(t)) rather than just rho(t) in the equation.

> tf := fsolve(eval(rho(t)) = L, t=1.5..2);

$$tf := 1.780597836$$

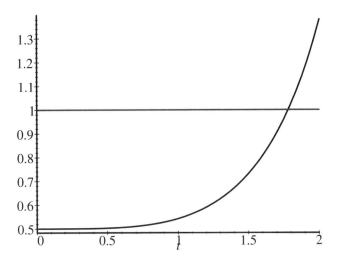

FIGURE 3.8. Radial position of the bead on the rotating wire as a function of time. The horizontal line represents the length of the rod.

```
> subs(t=tf, phi);
```
$$1.585264327$$

That is, the rod is just past vertical ($\frac{\pi}{2} \simeq 1.570796327$) when the bead leaves the end. Since the bead's speed is given by

$$v = \sqrt{\dot{\rho}^2 + \rho^2 \dot{\phi}^2} \,,$$

we can discover how fast it is moving when it leaves the end of the rod with

```
> sqrt( diff(rho(t),t)^2 + (rho(t)*diff(phi,t))^2 ):
> evalf(subs(t = tf, "));
```
$$2.216223190$$

The units are m/s. Of course, once the bead leaves the rod Eq. 3.20 no longer applies; it then behaves as a projectile, subject only to the gravitational force.

3.4 Work and Energy

3.4.1 *Projectile Motion from an Energy Viewpoint*

Let us imagine launching a projectile with a speed v_0 from a height y_0 above the ground. We choose a coordinate system with the origin at ground level, the x-axis horizontal, and the y-axis vertical. The zero potential energy level is also taken at ground level, so that $U = mgy$. Neglecting air resistance, conservation of mechanical energy yields

$$\frac{1}{2}mv^2 + mgy = \frac{1}{2}mv_0^2 + mgy_0 \; .$$

The right side of this equation is the initial energy, while the left side is the energy at some arbitrary point later in the flight. After the object hits the ground, of course, the equation no longer applies. This equation can be used to determine the speed of the particle at a given height y, or the height(s) at which the projectile reaches a given speed. The mass of the projectile cancels, and so has no effect on the motion.

Since the x-component of the velocity does not change, the equation can be simplified to obtain

$$\frac{1}{2}v_y^2 + gy = \frac{1}{2}v_{0y}^2 + gy_0 \; . \tag{3.21}$$

We can use Eq. 3.21 to find the maximum height by setting $v_y = 0$ and solving the resulting equation for y. Doing so yields

$$y_{max} = \frac{v_{0y}^2}{2g} + y_0 \; .$$

If we solve Eq. 3.21 for v_y, we find that

$$v_y = \pm\sqrt{v_{0y}^2 + 2g(y_0 - y)} \; . \tag{3.22}$$

The positive root applies to the upward portion of the projectile's path, while the negative root applies to the downward portion. Furthermore, the physical situation limits y to the range $0 \le y \le y_{max}$. Equation 3.22 indicates that at each height $y \ge y_0$ the magnitude of the y-component of the velocity is the same on the way up as on the way down. From this it can be inferred that the time it takes the object to rise from a given height y to its maximum height is the same as the time it takes it to fall from the maximum height back to the height y, independent of the x-component of the velocity.

By using the fact that $v_y = dy/dt$ in Eq. 3.22, separating the y and t parts, and integrating over time from 0 to t, we obtain the following expression:

$$\int_{y_0}^{y} \frac{dy'}{\sqrt{v_{0y}^2 + 2g(y_0 - y')}} = t \; . \tag{3.23}$$

This result applies only to the rising part of the motion since we used the positive root of Eq. 3.22 to derive it. It can be used to find, e.g., the time to the maximum height (t_{max}) by replacing the upper limit on the y' integral by y_{max}. A similar equation for the falling portion of the path can be generated by replacing the lower limits on the y and t integrations by y_{max} and t_{max}, respectively, and then using the negative root from Eq. 3.22.

Equation 3.23 can be integrated, and the resulting equation solved for y as a function of t. These steps are particularly easily accomplished with Maple.

```
> eq := int(1/sqrt(v0y^2+2*g*(y0-yp)), yp=y0..y) = t:
> solve(eq, y);
```

$$y0 - v0y\, t - \frac{1}{2}t^2 g,\ y0 + v0y\, t - \frac{1}{2}t^2 g$$

The second solution is the applicable one since we require $dy/dt = v_{0y}$ at $t = 0$. It of course agrees with the usual expression for $y(t)$ for a projectile.

Note that the approach of using the conservation of mechanical energy equation to solve for velocity, and then integrating to find the time as a function of position analogous to Eq. 3.23 can be applied to any one-dimensional motion in which only conservative forces do work, provided care is taken in the choice of the positive or negative root for the velocity. However, in general we may not be able to analytically perform the position integration, or to invert the time-as-a-function-of-position equation to obtain position as a function of time. Nevertheless, the approach can still be quite useful, as when it is used to obtain corrections to the period of a simple pendulum in Chapter 4.

An alternative to this approach is to solve the differential equation, Eq. 3.22, directly. That is,

```
> eq := diff(y(t), t) = sqrt(v0y^2 + 2*g*(y0 - y(t)) ):
> dsolve({eq, y(0)=y0, D(y)(0)=v0y}, y(t)):
> simplify(");
```

$$y0 + v0y\, t - \frac{1}{2}t^2 g$$

Note that we need to specify the initial condition on dy/dt to insure that Maple chooses the correct sign for the result of the sqrt function. This approach has an advantage over the previous one because for more general cases that do not have analytical solutions it is generally easier to solve the differential equation numerically than to do the required integral over y and then invert to get $y(t)$. When applicable, the energy approach is quite useful in that it allows us to do explicitly one integration over time, thus reducing the problem to a first-order differential equation. This can be particularly helpful for problems that are not analytically solvable.

3.4.2   An Electrostatic Example

For our next example we evaluate the potential energy of and force on a negative point charge $-q$ in the vicinity of two fixed positive point charges of magnitude Q. The positive charges are separated by a distance of a. With an appropriate choice of units, this could represent an electron in an H_2^+ molecular ion. The situation is pictured in Fig. 3.9, although the z-axis is not shown

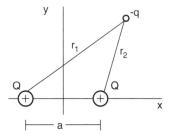

FIGURE 3.9. Point charge $-q$ in the neighborhood of two fixed point charges.

there. Noting that $r_1 = \sqrt{(x + \frac{a}{2})^2 + y^2 + z^2}$ and $r_2 = \sqrt{(x - \frac{a}{2})^2 + y^2 + z^2}$, we use the superposition principle to write the potential energy of the negative charge as the sum of the contributions due to interactions with the two positive charges:

$$U = -\frac{qQ}{4\pi\epsilon_0}\left(\frac{1}{r_1} + \frac{1}{r_2}\right).$$

The force on $-q$ can be obtained by computing the negative gradient of the potential energy. If we have defined r_1, r_2, and U as above, we can find the components of the force with Maple's diff command. For example,

```
> Fx := -diff(U, x):
```

and similarly for Fy and Fz. Alternatively, we can use Maple's grad command directly to evaluate the gradient. The grad command is in the linear algebra (linalg) package. We can load the package using with, or employ the long form for the function call. The former method is more convenient here so that we can use the curl command as well. Note that the warnings issued after the with command inform us that the commands norm and trace now have new definitions.

```
> with(linalg):
```

Warning new definition for norm
Warning new definition for trace

```
> F := -grad(U, [x,y,z]);
```

This latter statement returns the force as a list of x, y, and z components. With some hand simplification, the components of F are

$$F_x = -\frac{qQ}{4\pi\epsilon_0}\left(\frac{x + \frac{a}{2}}{r_1^3} + \frac{x - \frac{a}{2}}{r_2^3}\right),$$

$$F_y = -\frac{qQ}{4\pi\epsilon_0}\left(\frac{y}{r_1^3} + \frac{y}{r_2^3}\right),$$

and

$$F_z = -\frac{qQ}{4\pi\epsilon_0}\left(\frac{z}{r_1^3} + \frac{z}{r_2^3}\right).$$

A check on the results is provided by verifying that the curl of the force is zero, as it must be for a conservative force.

> curl(F, [x,y,z]);

$$[0\ 0\ 0]$$

For this problem the potential energy surfaces are axially symmetric about the line joining the two positive charges (x-axis). Consequently, all planes containing that axis have identical potential energy contours. This means we can choose $z = 0$, for example, and look at the x- and positive y-dependence only. In addition, there is a reflection symmetry through the $x = 0$ plane, so that for any constant C, the contour lines for the $x = C$ plane are duplicated by those of the $x = -C$ plane.

To help visualize the possible motion of the negative charge, it is useful to plot the potential energy as a function of x and y for $z = 0$. For such plotting, it is convenient to choose the unit of length to be a, and the unit of energy to be $qQ/(4\pi\epsilon_0 a)$. These choices effectively correspond to the following assignments.

> a := 1:
> U := unapply(-1/r1 - 1/r2, x, y, z):

The potential energy is defined as a function for convenience in examining different planes.

Our first thought is to use a three-dimensional plot to examine the dependence of U(x,y,0) on x and y. Unfortunately, plot3d is not very useful in this context because it does not have a provision for limiting the range of the function plotted. The resulting plot appears nearly flat except for very sharp negative peaks at the sites of the charges, giving us little insight into the energy landscape. The simple plot command does allow us to limit the range of the plot and thus better see the variation of U(x,y,0). Hence, we first use plot to graph U(x,y,0) as a function of x for several values of y, restricting U(x,y,0) to be greater than -10.

> plot({U(x,0,0), U(x,0.2,0), U(x,0.5,0), U(x,1,0)}, x=-1..1, -10..0);

The result is shown in Fig. 3.10.

It is perhaps more revealing, however, to display the potential energy with a series of contour plots in various planes, keeping in mind the symmetries

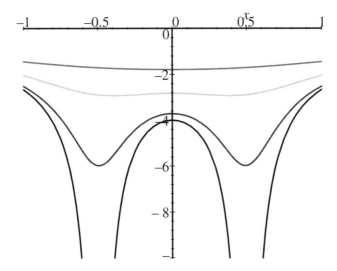

FIGURE 3.10. Potential energy $U(x, y)$ of a negative charge in the neighborhood of two positive charges, as a function of x. The four curves represent different y-values.

of the equipotential surfaces. These plots can be created with the Maple contourplot command, which is found in the plots package.

We first look at the energy contours in the x-y plane. As before there are possible problems with this because of the singularities at the sites of the two positive point charges. Fortunately, Maple's view option to contourplot avoids these difficulties by providing bounds for the minimum and maximum values to be plotted. Thus, the potential energy contours in the x-y plane are shown with the Maple statements

```
> with(plots):
> contourplot(U(x,y,0), x=-1..1, y=-1..1, view=-10..0, grid=[100,100]);
```

The result is shown in Fig. 3.11. The 100×100 grid chosen is a compromise between computational time and visual appearance.

Since the magnitude of the force at any point is equal to that of the gradient of the potential energy, the force is larger in regions where the equipotential contour lines are closer together. The direction is perpendicular to the equipotential contours, pointing to lower values of the potential energy. Also, since the kinetic energy is positive semi-definite, the particle is allowed to move only in those regions of the contour plot where the potential energy is less than the total energy. Thus contour plots are a useful aid toward visualizing the force that the particle experiences, and the region to which its motion is confined.

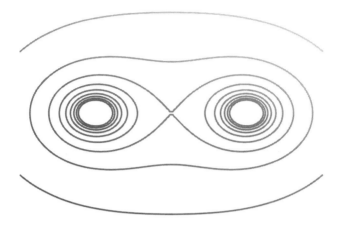

FIGURE 3.11. The potential energy contours of a negative charge in the neighborhood of two positive charges separated by a fixed distance.

What will the equipotential contours in the y-z plane look like? Try plotting them to verify your expectations. Check the statement about axial symmetry by comparing the $y = 0$ (x-z plane) contour plot with that for $z = 0$ (x-y plane).

We can plot the vector force field in a given plane using the Maple fieldplot command, provided that we have the in-plane components of the force. However, since the force is the negative gradient of potential energy, in this case it is easier to use the gradplot command. Unfortunately, there is currently no way in Maple to exclude a region or magnitude from a fieldplot or gradplot, so we must manually arrange to exclude the location of the two point charges. One way of doing this is look at the gradplot in a plane slightly above that in which the two charges lie: *e.g.*, $z = 0.01$. To illustrate, the result for the force field when the negative charge lies in the $z = 0.01$ plane is given in Fig. 3.12. This is obtained with the Maple command

```
> gradplot(-U(x,y,0.01), x=-1..1, y=-1..1, grid=[25,25],
>      axes=none, arrows=SLIM);
```

With this plot, and similar ones in other planes, we can visualize the "lines of force," in analogy with the electric field lines that are commonly shown in the second introductory physics course. The combination of such plots with the potential energy contour plots provides a powerful tool for the visualization of the dynamics of a given problem.

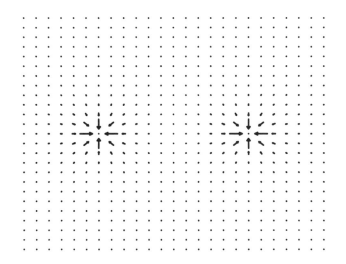

FIGURE 3.12. The force field in a plane slightly offset from the line joining two fixed positive charges.

3.5   Power

For some problems we are interested not only in the work done, but also in how rapidly it is done. The rate at which work is done by a given force is the *power* produced or absorbed by that force,

$$P = \frac{dW}{dt} \, .$$

Because of the relationship between work and potential energy for conservative forces, and between the work done by the net force and the change in kinetic energy, power can sometimes be thought of as the rate that energy is added to or removed from an object.

Since the work done by force $\mathbf{F}$ as an object's position changes differentially by $d\mathbf{r}$ is $dW = \mathbf{F} \cdot d\mathbf{r}$, the power produced or absorbed by $\mathbf{F}$ can be written as

$$P = \mathbf{F} \cdot \frac{d\mathbf{r}}{dt} = \mathbf{F} \cdot \mathbf{v} \, . \tag{3.24}$$

If the force used in Eq. 3.24 is the net force, we may apply Newton's second law to replace $\mathbf{F}$:

$$P_{net} = m\frac{d\mathbf{v}}{dt} \cdot \mathbf{v} = m\frac{d}{dt}\left(\frac{1}{2}v^2\right) \, .$$

Thus the power produced by the net force is equal to the rate of change of the kinetic energy:

$$P_{net} = \frac{dK}{dt} \, . \tag{3.25}$$

3.5.1   Object Falling Through a Viscous Fluid

To illustrate these ideas, we consider again a problem from Sec. 3.3.1, the example of projectile motion with air resistance. As noted there, the calculated y-motion also applies to an object falling through a viscous fluid. The two forces acting are the gravitational force $\mathbf{F}_{grav} = -mg\hat{y}$, where the positive y-direction is upward, and the drag force $\mathbf{F}_{drag} = -b\mathbf{v}$. The solution for the velocity, assuming it starts from rest, is given by Eq. 3.18 with $v_{0y} = 0$; that is,

$$\mathbf{v} = -\frac{mg}{b}\left(1 - e^{-\frac{bt}{m}}\right)\hat{y} \,. \tag{3.26}$$

From this the power produced by the gravitational force, and that absorbed by the drag force are obtained as functions of time:

$$P_{grav} = -mg\hat{y} \cdot \mathbf{v} = \frac{m^2 g^2}{b}\left(1 - e^{-\frac{bt}{m}}\right)$$

$$P_{drag} = -b\mathbf{v} \cdot \mathbf{v} = -\frac{m^2 g^2}{b}\left(1 - e^{-\frac{bt}{m}}\right)^2 \,.$$

These equations show that P_{grav} starts at zero and increases toward a positive constant value as the speed approaches that of the terminal speed, mg/b. P_{drag} on the other hand, starts at zero and decreases as energy is removed from the object by the viscous friction. P_{drag} approaches a negative constant value as the speed nears the terminal speed. It is easily seen that $P_{drag} = -P_{grav}$ in the long-time limit.

These predictions can be investigated with Maple. We obtain the speed of the particle by taking the absolute value of the VY(b,t) function for the y-component of the velocity in Section 3.3.1. We further define Pgrav, Pdrag, and Pnet with the Maple function definitions

```
> Pgrav := (b,t) -> m*g*abs(VY(b,t)):
> Pdrag := (b,t) -> -b*VY(b,t)^2:
> Pnet := (b,t) -> Pgrav(b,t) + Pdrag(b,t):
```

These definitions make sense only after going through the series of Maple commands in Sec. 3.3.1 that led to the definition for VY(b,t), of course. Now, we define a plotting function

```
> PL := (b,T) -> plot({Pgrav(b,t), Pdrag(b,t), Pnet(b,t)}, t=0..T):
```

and make the parameter assignments

```
> m:=1:   g := 9.8:   v0y := 0:
```

to display the power as a function of time. For example,

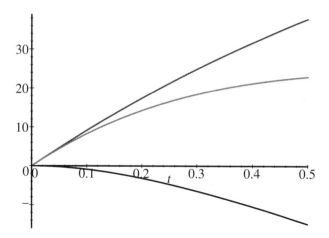

FIGURE 3.13. Short-time behavior of the power received and given up by an object falling through a viscous fluid.

> PL(1, .5);
> PL(1, 5);

yield plots similar to those shown in Figs. 3.13 and 3.14. Figure 3.13 shows P_{grav}, P_{drag}, and P_{net} for $b = 1\,kg/s$ over the time range 0 to 0.5 seconds. The predicted small-time increase in P_{grav} and decrease in P_{drag} are clearly seen. P_{drag} decreases more slowly than P_{grav} increases for small times (thus small v) since it is proportional to the square of the speed, while the latter increases linearly with v. Consequently, the power produced by the net force increases, and the rate at which the particle gains kinetic energy increases during this period.

Figure 3.14 shows the same three quantities over a longer time interval. In this figure the positive saturation of P_{grav} and the negative saturation of P_{drag} are clearly seen for $t \geq 4\,s$. The power produced by the net force, which initially increases at small times, begins to decrease, approaching zero at longer times, in agreement with the long-time limit of $P_{drag} = -P_{grav}$. This is also predicted by Eq. 3.25 since the kinetic energy approaches a constant value as the speed approaches the terminal speed.

3.6 Angular Momentum and Torque

The *angular momentum* of a particle about the origin is defined by the equation

$$\mathbf{L} \equiv \mathbf{r} \times \mathbf{p},$$

where $\mathbf{r}$ is the position vector of the particle, and $\mathbf{p}$ is its momentum. As illustrated in the example in the next section, both the magnitude and direction

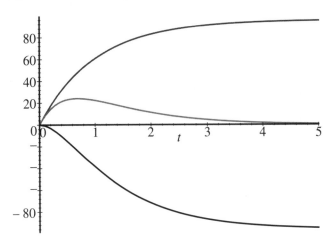

FIGURE 3.14. Long-time behavior of the power received and given up by an object falling through a viscous fluid.

of the angular momentum are dependent upon the choice of origin.

Taking the derivative of **L** with respect to time we find

$$\frac{d\mathbf{L}}{dt} = \frac{d\mathbf{r}}{dt} \times \mathbf{p} + \mathbf{r} \times \frac{d\mathbf{p}}{dt} .$$

Since $d\mathbf{r}/dt$ is in the same direction as **p**, and $d\mathbf{p}/dt$ is equal to the net force acting on the particle, this equation reduces to

$$\frac{d\mathbf{L}}{dt} = \mathbf{r} \times \mathbf{F}_{net} .$$

This relationship provides the justification for defining the *torque* about the origin due to a force **F** acting on a particle as

$$\tau \equiv \mathbf{r} \times \mathbf{F} ,$$

which leads to the following relationship between the net torque on the particle and its angular momentum:

$$\tau_{net} = \frac{d\mathbf{L}}{dt} . \tag{3.27}$$

One consequence of this result is a law for *conservation of angular momentum*:

If the net torque acting on a particle is zero, the angular momentum of the particle is constant.

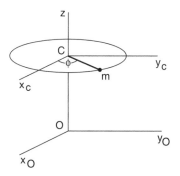

FIGURE 3.15. Schematic of a particle in circular motion with two reference systems.

3.6.1   Circular Motion

To illustrate that both **L** and τ depend upon the choice of origin, we consider a mass m which is undergoing uniform circular motion. This could be, *e.g.*, a satellite moving in circular orbit around the earth, or the mass of a conical pendulum. If we choose the origin at point C, the center of the circle, the motion is in the x-y plane. We use cylindrical coordinates and assume that the mass is moving in the direction of increasing ϕ. See Fig. 3.15.

Since the force responsible for uniform circular motion is toward the center, $\mathbf{r} \times \mathbf{F}$ vanishes; consequently, the torque about point C is zero, $\tau_C = 0$. The angular momentum is

$$\mathbf{L}_C = \rho \hat{\rho} \times mv \hat{\phi} = mv\rho \hat{z} .$$

For uniform circular motion this is constant in both magnitude and direction, as it should be since the torque vanishes.

If, however, the same problem is examined with the origin offset by a distance a from the force center (point O in the figure), the result for the angular momentum is

$$
\begin{aligned}
\mathbf{L}_O &= (\rho \hat{\rho} + a \hat{z}) \times mv \hat{\phi} \\
&= mv\rho \hat{z} - mva \hat{\rho} .
\end{aligned}
\tag{3.28}
$$

In this case, the angular momentum consists of a constant component (L_z) and a component whose magnitude is constant but whose direction changes with the rotation (L_ρ). If we take the time derivative of Eq. 3.28 and use Eq. 3.4 we find

$$\frac{d\mathbf{L}_O}{dt} = -mva\dot{\phi} \hat{\phi} .$$

The torque due to the attractive force **F** whose force center is offset from the origin of the coordinate system by $a \hat{z}$ is

$$
\begin{aligned}
\tau_O &= (\rho \hat{\rho} + a \hat{z}) \times (-F \hat{\rho}) \\
&= -aF \hat{\phi} .
\end{aligned}
\tag{3.29}
$$

Since for circular motion $F = mv^2/\rho = mv\dot{\phi}$, this case also satisfies the expected relationship between torque and angular momentum,

$$\tau_O = \frac{d\mathbf{L}_O}{dt} ,$$

even though τ_O and $\mathbf{L}_O$ are different from τ_C and $\mathbf{L}_C$. An interesting aside is the fact that $\mathbf{L}_O$, when averaged over a full rotational period, yields $\mathbf{L}_C$.

The concepts of torque and angular momentum in classical mechanics are most useful when dealing with a system of particles, such as a rigid body, rather than a single particle, where we might as well use Newton's 2nd law. However, angular momentum, particularly in the form of the conservation law suggested by Eq. 3.27 when $\tau_{net} = 0$, plays an important role in quantum mechanics. This is true despite the fact that particles cannot simultaneously have precisely defined position and momentum.

3.7   Central Forces

A central force is one whose magnitude depends only on the distance of a particle from a single point and whose direction is along the line joining that point and the particle. The force on the particle can be attractive (directed toward the force center) or repulsive (directed away from the center). If we take the origin of the coordinate system to be at the force center, these are the $-\hat{\mathbf{r}}$ and $\hat{\mathbf{r}}$ directions, respectively, in spherical coordinates. The force then has the form

$$\mathbf{F(r)} = f(r)\hat{\mathbf{r}} , \qquad (3.30)$$

where $f(r)$ depends only on the magnitude of $\mathbf{r}$.

The torque on the particle due to central forces is zero when taken about the origin since $\mathbf{r} \times \hat{\mathbf{r}} = 0$. Consequently, the angular momentum of the particle taken about the origin is constant. This leads to the result that a particle moving solely under the influence of a central force remains in the plane defined by the vectors $\mathbf{r}$ and $\mathbf{p}$. If the direction of $\mathbf{L}$ is taken to be the z-axis, we can use cylindrical coordinates to describe central-force motion, with the particle remaining in the (ρ, ϕ) plane. The force is then in the $\pm\hat{\rho}$ direction.

3.7.1   Motional Constants Under Central Forces

Since the particle remains in the (ρ, ϕ) plane, its velocity can be written as in Eq. 3.5. For clarity, we replace ρ by r in that expression; they are identical in this case since $z = 0$. Thus, using this expression for the velocity, the angular momentum for a particle in a central force is

$$\mathbf{L} = \mathbf{r} \times m\mathbf{v} = mr^2\dot{\phi}\hat{\mathbf{z}} . \qquad (3.31)$$

Since the torque due to a central force about the force center is zero, angular momentum is a constant of the motion. Furthermore, as shown by most standard classical mechanics texts, the rate at which the area swept by the position vector to the particle changes in time is given by

$$\frac{dA}{dt} = \frac{1}{2}r^2\dot{\phi} = \frac{L_z}{2m} .$$

Thus Kepler's law regarding equal areas swept in equal times is a consequence of the fact that angular momentum is conserved by a central force.

Furthermore, since the work done by a central force as the particle moves from one point to another depends only on the radial components of the initial and final position,

$$\int_{\mathbf{r}_0}^{\mathbf{r}} \mathbf{F}(\mathbf{r}') \cdot d\mathbf{r}' = \int_{r_0}^{r} f(r') \, dr' ,$$

central forces are conservative. Thus, the sum of kinetic and potential energies is constant for motion in a central force. That is,

$$E = \frac{1}{2}mv^2 + U(r) , \tag{3.32}$$

is a constant of the motion. $U(r)$ is the potential energy function, which depends only upon the radial coordinate r and the reference point r_0:

$$U(r) = U(r_0) - \int_{r_0}^{r} f(r') \, dr' .$$

3.7.2 Calculation of Orbits

As an example, we calculate the orbital equation for a particle moving in a central force. To be specific, we consider a particle under the influence of a force which varies inversely as the square of the distance from the force center. In terms of the distance r the potential energy is

$$U(r) = -\frac{k}{r} .$$

The gradient of this expression yields a $1/r^2$ force, which is attractive for positive k, and repulsive for negative k. This potential energy applies, e.g., to the Kepler problem if $k = Gm_1m_2$ or the Rutherford problem if $k = q_1q_2/(4\pi\epsilon_0)$.

If we use the expression for velocity in polar coordinates, Eq. 3.5, we can write the (constant) total energy of the particle as

$$
\begin{aligned}
E &= \tfrac{1}{2}mv^2 + U(r) \\
&= \tfrac{1}{2}m(\dot{r}^2 + r^2\dot{\phi}^2) - \frac{k}{r} .
\end{aligned}
$$

We solve this equation to find the orbit. In doing so, we use

$$\dot{\phi} = \frac{L}{mr^2},$$

obtained from the expression for angular momentum, Eq. 3.31.

The solution for the orbit equation proceeds as follows. We first employ Maple to obtain an expression for the kinetic energy in terms of $r(\phi)$. The alias command simplifies the entry of the expression. (See ?alias for information.)

```
> alias(phi = phi(t)):    alias(r = r(phi(t))):
> 1/2*m*(diff(r,t)^2 + r^2*diff(phi,t)^2):        # kinetic energy
> subs(diff(phi,t)=L/(m*r^2), "):
> factor(");
```

$$\frac{1}{2} \frac{L^2 \left(D(r)(\phi)^2 + r^2\right)}{mr^4}$$

The result is an expression for the kinetic energy, given in terms of Maple's differential operator D. The notation $D(r)(\phi)$ that Maple returns can be understood as $dr/d\phi$.

By changing the aliases for ϕ and r we can apply this expression for kinetic energy to the energy conservation equation and solve it for $r(\phi)$:

```
> alias(phi = 'phi'): alias(r = r(phi)):
> eq := L^2*(diff(r,phi)^2 + r^2)/(2*m*r^4) - k/r = En:
> sol := dsolve(eq, r);
```

$$sol \quad := \quad -\frac{arctanh\left(\frac{1}{2}\frac{-2L^2+2kmr}{\sqrt{-L^2}\sqrt{-L^2+2kmr+2En\,mr^2}}\right) L}{\sqrt{-L^2}} + \phi = \_C1,$$

$$\frac{arctanh\left(\frac{1}{2}\frac{-2L^2+2kmr}{\sqrt{-L^2}\sqrt{-L^2+2kmr+2En\,mr^2}}\right) L}{\sqrt{-L^2}} + \phi = \_C1$$

These solutions are not valid for the special case $L = 0$, which corresponds to a head-on collision.

Maple cannot explicitly solve the orbital differential equation for $r(\phi)$. However, it is able to return two implicit solutions, each in the form of ϕ plus an expression involving r equal to an arbitrary constant. Since the constant is additive to ϕ, it can be set to zero by an appropriate choice of the axis from which ϕ is measured. Its value does not affect the shape of the orbit, only its orientation. With the constant set to zero we can see that the resulting solutions for ϕ in terms of r are the negative of each other, corresponding to movement around the orbit in clockwise and counterclockwise directions. Since we are interested in the shape of the orbit, we can choose either of the solutions. To simplify the result we use the 'symbolic' option to simplify since all the variables are real. Thus, the equation is simplified by the Maple commands

```
> _C1 := 0:  simplify(sol[1], symbolic);
```

$$\arctan\left(\frac{-L^2 + kmr}{L\sqrt{-L^2 + 2kmr + 2En\,mr^2}}\right) + \phi = 0$$

Maple cannot solve this equation for $r(\phi)$ either. Unfortunately, due to the implicit ϕ dependence in r due to our use of alias, it cannot even solve it for ϕ. However, we can force the latter solution by rearranging the equation with the command

```
> phi = -(lhs(") - phi);
```

$$\phi = -\arctan\left(\frac{-L^2 + kmr}{L\sqrt{-L^2 + 2kmr + 2En\,mr^2}}\right)$$

Some experimentation leads to the discovery that if we take the sine of each side of this equation we get an equation that Maple *can* solve for $r(\phi)$.

```
> sin(phi) = simplify(sin(rhs(")), symbolic);
```

$$\sin(\phi) = -\frac{-L^2 + kmr}{\sqrt{mr}\sqrt{2L^2\,En + k^2m}}$$

```
> solve(", r);
```

$$\frac{L^2}{\sin(\phi)\sqrt{m}\sqrt{2L^2\,En + k^2m} + km}$$

Minor manual manipulation of this result yields

$$r(\phi) = \frac{L^2/(km)}{1 + \sqrt{1 + 2EL^2/(k^2m)}\,\sin\phi} \tag{3.33}$$

in terms of the original symbol E for energy.

Equation 3.33 is the polar-coordinate equation for a conic section. The eccentricity ϵ can be identified as the coefficient of $\sin\phi$ in the denominator. The value of ϵ determines the kind of orbit: circle ($\epsilon = 0$), ellipse ($0 < \epsilon < 1$), parabola ($\epsilon = 1$), or hyperbola ($\epsilon > 1$). For an attractive force ($k > 0$), any of these orbits is possible, depending on the values of E. Negative total energy results in circular or elliptical orbits, and the motion is periodic and bounded. Zero total energy yields a parabolic orbit, and positive energy a hyperbolic path. In either of these latter cases the particle is not bound, and approaches an asymptote defined by $\sin\phi = -1/\epsilon$. For the hyperbolic orbit case there are two branches. One of these corresponds to an attractive force, while the other is appropriate for a repulsive force.

Suppose we repeat the calculation with an additional δ/r^2 term added to the potential energy. Such a term can arise, *e.g.*, in the potential energy of a point charge moving about fixed charge distribution which is not spherically symmetric. The result for the orbit of such a system is

$$r(\phi) = \frac{(L^2 + \delta m)/(km)}{1 + \sqrt{1 + 2E(L^2 + \delta m)/(k^2 m)} \sin(\sqrt{1 + \delta m/L^2}\,\phi)} \, . \tag{3.34}$$

Comparison of this result with Eq. 3.33 reveals that it is almost of the form of a conic section. If $E < 0$, we can in fact interpret Eq. 3.34 as the equation for a precessing ellipse, since for small $|\delta|$ the particle returns to the value of $r(0)$ slightly before $\phi = 2\pi$ (for $\delta > 0$) or slightly after $\phi = 2\pi$ (for $\delta < 0$).

Show that Eq. 3.34 is a solution to the orbit equation for a force that has both $1/r^2$ and $1/r^3$ terms.

3.8   Problems

As before, "verify" means not only obtain the requested limit, but also provide a physical argument for its correctness.

1. A free electron in the ionosphere is subjected to a sinusoidally-varying electric field, $\mathbf{E} = E_0 \cos(\omega t + \phi)\,\hat{\mathbf{x}}$, as an incoming electromagnetic wave passes. (Ignore the accompanying magnetic field.)

    (a) Solve for the position and velocity of the electron as a function of time, assuming that it starts from rest at the origin.

    (b) Show that for general values of the phase ϕ the velocity of the electron averaged over the period of the field is non-zero, so that the electron drifts in the $\pm\hat{\mathbf{x}}$ direction, depending on the value of ϕ.

    (c) Explain physically the motion that occurs for the special cases $\phi = 0$ and $\phi = \frac{\pi}{2}$.

2. A rope of mass m and length L is placed on a table with a small piece of length $L_0 < L$ hanging over the edge. It begins to slide off, falling under the influence of the earth's gravity.

    (a) Obtain an expression for the position of the lower end of the rope as a function of time for $0 \le t \le T$, where T is the time at which the upper end of the rope leaves the table. Express the result in terms of hyperbolic trigonometric functions.

    (b) Find T in terms of the given quantities m, L, L_0, and g.

    (c) How fast is the rope moving when it leaves the table? Verify the $L_0 \to L$ limit. How does the $L_0 \to 0$ limit compare with the speed the rope would have if were it held vertically with the lower end at the level of the table top and then allowed to fall its length?

3. A particle of charge q moves with a velocity $v = v_0\hat{\mathbf{x}}$ into a region of constant electric and magnetic fields, $\mathbf{E} = E_0\hat{\mathbf{y}}$ and $\mathbf{B} = B_0\hat{\mathbf{z}}$. In this region it experiences a Lorentz force, $\mathbf{F} = q(\mathbf{E} + \mathbf{v} \times \mathbf{B})$.

    (a) Write down the equations of motion for the velocity components in the x-, y-, and z-directions. Note that the particle is not accelerated in the z-direction, and so remains in the x-y plane.

    (b) Solve the coupled equations for $v_x(t)$ and $v_y(t)$.

    (c) Show that the particle undergoes uniform circular motion for the special case $E_0 = 0$.

    (d) Note that if the particle starts at rest it obtains a non-zero drift velocity in the x-direction *even though the electric field is in the y-direction*. Explain physically why this occurs.

(e) Show that there is a particular value for v_0 such that the particle continues to move at a constant velocity through the crossed **E** and **B** fields.

4. Suppose a car engine delivers a constant power P to the vehicle when the accelerator is fully depressed.

   (a) Neglecting air resistance, find the force on the car and solve the equation of motion for $v(t)$. Is there a terminal speed?

   (b) Assume that air resistance is proportional to the velocity, $F_{air} = -bv(t)$. Solve the equation of motion for $v(t)$. What is the terminal speed in this case?

   (c) For a "typical" small car, $m = 1000$ kg, $P = 7.46 \times 10^4$ W (100 hp), and $b = 40$ kg/s. Plot the velocity as a function of time, assuming that it starts from rest and accelerates at the maximum rate.

   (d) Plot the acceleration of the car as a function of time, again assuming the car starts from rest. What does the small-time behavior suggest about the feasibility of the engine supplying constant power to move the car?

5. Consider an object falling from rest under the influence of gravity and a drag force.

   (a) Solve the equation of motion for the velocity for two different drag force magnitudes, $F_{drag} = \alpha v$ and $F_{drag} = \beta v^2$. In each case the drag force should be in the opposite direction to the velocity.

   (b) Find the values of α and β required to give the same terminal speed, v_∞.

   (c) Assign the above values to α and β and make a small-time expansion of the two $v(t)$ cases. Which deviates more quickly from the linear increase in speed characteristic of the no-drag case? Why?

   (d) Assign values $m = 1$ kg, $g = 9.8$ m/s^2, and $v_\infty = 20$ m/s and plot the two cases as functions of time on the same plot. Which case approaches the terminal speed more quickly? Why?

6. The resistive force on a sphere of radius r moving with speed v may be written approximately as

$$f = \alpha r v + \beta r^2 v^2 .$$

Its direction, of course, is opposite to the direction of the velocity.

   (a) Solve the equation of motion for the speed of a spherical particle falling from rest in the earth's gravitational field. Express the mass in terms of the density, ρ, and radius.

(b) For motion through the atmosphere at a wide range of speeds the constants may be taken as $\alpha = 0.00031$ N·s/m^2 and $\beta = .87$ N·s^2/m^4[8]. Using the density of water, $\rho = 1000$ kg/m^3 and $g = 9.8$ m/s^2, find the terminal speed of a raindrop of radius 2.0 mm.

(c) Determine the time it takes for the raindrop to reach 99% of its terminal speed, starting from rest. Find the distance the drop has fallen in this amount of time. How does this distance compare with your height?

7. A particle of mass m is attracted by gravity to a spherical planet of radius R and mass M.

(a) Find the time required for it to hit the planet's surface if it starts from rest at a height h above the surface.

(b) Show that for small h ($h \ll R$) the time is approximately that expected for falling in a uniform gravitational field of magnitude g, where $g = GM/R^2$.

4

The Harmonic Oscillator

4.1 Linear Restoring Force

As an extended example of Newtonian dynamics, let us examine the motion of a particle under the influence of a force whose magnitude is proportional to the displacement of the particle from an equilibrium position, and whose direction is toward that equilibrium position. If we choose the coordinate system so that the origin is at the point of equilibrium, the force may be written

$$\mathbf{F} = -k\mathbf{r} , \tag{4.1}$$

where k is a positive constant. This linear restoring force is often referred to as a Hooke's law force, after Robert Hooke, who first observed that the deformation of a solid (such as the stretch of a spring or the bending of a bar) is resisted by a force whose magnitude is approximately proportional to the deformation, as long as the deformation is not so large as to exceed the elastic limit. It is one of the most important forces in the study of physics because it provides a good approximation to many real physical systems and because its equations of motion in both classical and quantum mechanics have analytic solutions.

The linear restoring force is a central force and, as such, is conservative. The potential energy function is given by $U = \frac{1}{2}kr^2$. Thus, a particle under the influence of a net force of the form of Eq. 4.1 moves with constant mechanical energy; that is, $E = \frac{1}{2}mv^2 + \frac{1}{2}kr^2$ is constant. As noted in Chapter 3, the angular momentum about the equilibrium position is also constant, and the motion will be planar.

If the net force on a particle is given by Eq. 4.1, Newton's second law provides the equation of motion,

$$-k\mathbf{r} = m\ddot{\mathbf{r}} .$$

Applying rectangular coordinates, this equation can be written as three independent equations, corresponding to motion in the x-, y-, and z-directions. Some rearrangement yields

$$\ddot{x} + \frac{k}{m}x = 0 , \qquad (4.2)$$

with similar equations for y and z. The fact that the three-dimensional equation of motion separates into three independent one-dimensional equations of identical mathematical form means that we can confine our attentions to the one-dimensional problem; the three-dimensional result can be easily generated from it.

These equations depend on k and m only through their ratio, and it is worthwhile to emphasize this restricted dependence with a single parameter. Since both k and m are positive, the ratio is also positive. Thus, we write the x-equation as

$$\ddot{x} + \omega_0^2 x = 0 , \qquad (4.3)$$

where

$$\omega_0 = \sqrt{\frac{k}{m}} . \qquad (4.4)$$

The y- and z-equations can be obtained simply by replacing x by y and z, respectively. The motion described by Eq. 4.3 is known as *simple harmonic motion*. We examine it and some variations on it in the remaining sections of this chapter. Part of this chapter has been adapted from work previously published by the author.[9]

4.2   Simple Harmonic Motion

Many dynamical systems approximately satisfy Eq. 4.3. For the purposes of this section, however, let us think of it as the equation of motion for a mass m at the end of an ideal spring of spring constant k, confined to motion along the x-axis. The origin has been chosen so that the equilibrium position is at $x = 0$.

The parameter ω_0 represents an angular frequency. The mass m oscillates between positive and negative values of x with a frequency of $\frac{\omega_0}{2\pi}$, the *natural frequency* of oscillation for the simple harmonic oscillator. (The angular frequency ω_0 is also often referred to as the natural frequency.) As we will see, the natural frequency plays a very important role in determining the qualitative behavior of the harmonic oscillator.

The solution to Eq. 4.3, including the initial values, can be obtained with Maple. For example, the commands

```
> eq := diff(x(t),t,t) + omega0^2*x(t) = 0:
> sol := dsolve({eq, x(0)=x0, D(x)(0)=v0}, x(t));
```

yield the following solution for $x(t)$.

$$x(t) = x_0 \cos(\omega_0 t) + \frac{v_0 \sin(\omega_0 t)}{\omega_0} . \qquad (4.5)$$

As noted earlier, the linear restoring force is conservative. In one-dimension its associated potential energy function is $U = \frac{1}{2}kx^2$. We can verify the constancy of the mechanical energy with the Maple statements

```
> assign(sol):
> 1/2*m*diff(x(t),t)^2 + 1/2*k*x(t)^2:      # mechanical energy
> simplify(subs(omega0=sqrt(k/m), "));
```

The resulting expression for the energy is

$$E = \frac{1}{2}k x_0^2 + \frac{1}{2}mv_0^2 . \qquad (4.6)$$

Since x_0 and v_0 are constant, the result is constant energy, as expected.

4.2.1 Amplitude and Phase

An alternate form for the solution in Eq. 4.5 is

$$x(t) = A \cos(\omega_0 t + \theta) , \qquad (4.7)$$

where the amplitude A and phase θ depend on the initial values x_0 and v_0. This dependence is determined by equating the right sides of Eqs. 4.5 and 4.7 at two independent values of t, and solving the resulting two equations for A and θ. With manual calculation, judicious choice of the two values for t makes the calculation much simpler. For Maple, the choice is not so important, as long as we use values that Maple can treat exactly (*i.e.*, not floating-point), and the two values are not separated by an integral multiple of π/ω_0, insuring two independent equations. In this case, $t = 0$ and $t = \frac{\pi}{2\omega_0}$ do quite nicely, since $\sin \omega_0 t$ and $\cos \omega_0 t$ take on simple values at these two points. Recalling that we previously assigned x(t) the value obtained from solving the equation of motion, the Maple commands

```
> eq1 := A*cos(omega0*t + theta) = x(t):
> eq2 := subs(t=0, eq1):
> eq3 := subs(t=Pi/(2*omega0), eq1):
```

provide the equations that yield the amplitude and phase, A and θ. The obvious method is to ask Maple to solve them simultaneously.

> solve({eq2, eq3}, {A, theta});

$$\{\theta = 2\arctan\left(RootOf\left(v0\_Z^2 - 2x0\_Z\omega0 - v0\right)\right),$$

$$A = -\frac{v0 + x0\,RootOf\left(v0\_Z^2 - 2x0\_Z\omega0 - v0\right)\omega0}{\left(v0\_Z^2 - 2x0\_Z\omega0 - v0\right)\omega0}$$

Maple's solution is surprisingly complicated, and is not improved much by further manipulation. Computer algebra systems such as Maple can sometimes yield expressions that are not at all what we want, although they of course should be equivalent. We need to be aware of this trait and be willing to try other approaches when we do not get a helpful form for a solution. In this case we can get simpler expressions for A and θ by using the solve-and-substitute technique illustrated in Sec. 2.5.3. For example, we can first solve eq2 for A, assign the result, and then solve the resulting eq3 for θ.

> A := solve(eq2, A):
> theta := solve(eq3, theta):
> A;   theta;

yield simple values for A and θ:

$$A = x_0\sqrt{1 + \frac{v_0^2}{x_0^2\omega_0^2}} \tag{4.8}$$

and

$$\theta = -\arctan\left(\frac{v_0}{\omega_0 x_0}\right).$$

We can express the energy of the oscillator in terms of the amplitude by combining Eqs. 4.4, 4.6, and 4.8. The result is that $E = \frac{1}{2}kA^2$.

As the solution for $x(t)$ reveals, simple harmonic motion is steady state motion, endlessly repeating itself with a period of $T = \frac{2\pi}{\omega_0}$. The initial conditions determine the energy of the system, or equivalently, the amplitude of the oscillations, but do not affect the qualitative description of the motion. For any given values of x_0 and v_0, we get equivalent motion by starting the mass at rest with an initial position given by $x(0) = A$, as given in Eq. 4.8. The only effect is to shift the origin of time. Consequently, in further deliberations we often choose $v_0 = 0$ for convenience.

4.2.2 Scaling of the Equation of Motion

With the choice $v_0 = 0$, the solution for $x(t)$ has three free parameters — x_0, ω_0, and t. The parameter x_0 is equal to the amplitude of the oscillation. We

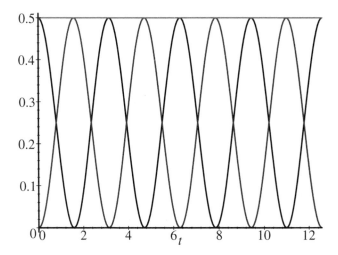

FIGURE 4.1. Kinetic, potential, and total energy as functions of time for the simple harmonic oscillator.

can set it equal to 1 and obtain results for other values by scaling lengths by the amplitude; that is, for $A \neq 1$ multiply $x(t)$ and $v_x(t)$ by A, energy by A^2, *etc.* In addition, since the solution for $x(t)$ depends on ω_0 and t only through the product $\omega_0 t$, we can take $\omega_0 = 1$. Results for other values of ω_0 are obtained from the $\omega_0 = 1$ results by changing the time scale appropriately. With these assignments for the parameters of the problem the position of the particle oscillates between ± 1 with a period of $T = 2\pi$. After making these assignments we can plot the kinetic and potential energies and their sum. Figure 4.1 shows the result, and it is obvious that although both the kinetic and potential energies vary regularly with time, the sum is constant.

Reproduce Fig. 4.1 by entering the appropriate Maple commands.

Before continuing the analysis of simple harmonic motion, we consider the scaling arguments outlined in the previous paragraph more formally. The primary purpose for scaling equations prior to solving them is to reduce the number of free parameters in the solutions, which simplifies analysis of the solution. The solution of the original equation for one-dimensional motion of a mass attached to a spring, Eq. 4.2, illustrates this point. The solution to Eq. 4.2 depends on both k and m. Had we not noticed that only the ratio of the two parameters is important, we would have had to plot the results for varying values of both k and m to be reasonably sure of understanding the qualitative behavior of the system. However, since the solution for the motion of a system with, for example, $m = 10$ kg and $k = 10$ N/m is exactly the same as that for $m = 1$ kg and $k = 1$ N/m, there is no reason to plot the two cases separately. Insight dictates that there is only one free parameter to consider, the ratio of k/m.

Scaling equations can reduce the number of free parameters. It can also reduce the range of the numbers involved in numerical or graphical calculations, which can make it easier to get accurate results for some problems. To illustrate the technique, let us first consider scaling of the independent variable (t) in Eq. 4.3. We do this by changing to a new variable, s, which is proportional to t, but has no units. That is, we change variables to $s = t/\tau$, where τ is a constant to be determined that has units of time. If we make this variable change and multiply both sides by τ^2, Eq. 4.3 becomes

$$\frac{d^2x}{ds^2} + \omega_0^2 \tau^2 x = 0 .$$

We are free to choose the value of the constant τ, so for simplicity take it to be $\tau = 1/\omega_0$. With this choice, the scaled simple harmonic oscillator equation becomes

$$\frac{d^2x}{ds^2} + x = 0 . \tag{4.9}$$

This equation has no free parameters! Its solution, for initial position and velocity of x_0 and v_0 is

$$x(s) = x_0 \cos(s) + v_0 \sin(s) . \tag{4.10}$$

Note, however, that this v_0 differs from that in Eq. 4.5 by a factor of ω_0, because "velocity" is now dx/ds rather than dx/dt.

To get Eqs. 4.9 and 4.10 we have, in effect, chosen a unit for time which is characteristic of the problem, in this case, $1/\omega_0$. Such characteristic units are sometimes called *natural units*. We can always recover the solution in terms of the original time units using the scaling equation $s = \omega_0 t$, making sure that any parameters whose values are expressed in terms of time units (such as v_0) are properly unscaled.

To clarify this further, consider starting the oscillator by pulling the mass away from equilibrium by some amount and letting it go from rest; thus $v_0 = 0$ and x_0 is the initial stretch of the spring. As noted earlier, the amplitude of the motion is equal to this initial stretch under these conditions. Equation 4.10 indicates that the mass oscillates with a period of 2π in the scaled time units, regardless of m and k. That is, all mass and spring constant combinations will demonstrate the same qualitative oscillatory behavior. It is not necessary to graph solutions for various ω_0 values to qualitatively understand the motion. The only quantitative difference for different k and m combinations is in the period of oscillations, which can be found by rescaling the 2π in s-units to $2\pi/\omega_0$ in t-units.

This simple example does not fully demonstrate the power of scaling equations to aid analysis of the motion; the analytic solution is simple enough to see that all m and k combinations oscillate sinusoidally. However, with more difficult problems the analytic solution may be too complex to comprehend the dependencies on the parameters, or there may not be an analytic solution at all. In these cases the ability to reduce the number of free parameters by scaling or other means can greatly simplify graphical analysis. Furthermore, scaling can also be of substantial benefit for numerical evaluation. Thoughtful scaling of the equations leads to a choice of units that are more appropriate in size for the calculated quantities, thus greatly reducing the chance that overflow or underflow errors will negatively affect the results.

We can often profitably scale the dependent variable as well. A linear differential equation is not changed by scaling the dependent variable, although the solution will be. For the simple harmonic oscillator, a natural unit of length is the amplitude of the oscillation, which is proportional to the square root of the energy. If we scale all distances in such a way that the scaled amplitude is one, the new version of Eq. 4.8 provides a relationship between x_0 and v_0 so that only one free parameter remains in the scaled solution, even for arbitrary initial conditions.

4.2.3   Phase Plots

Phase plots are another useful tool for the analysis of motion, particularly in one-dimension. In one-dimension, phase plots are graphs of velocity $vs.$ position, often for multiple values of the parameters of the system. For many problems it is possible to obtain an equation relating velocity to position even when there is no analytic solution for position as a function of time. From phase plots stable and unstable equilibrium positions can be found, conditions for periodic motion can be discovered, and the general behavior of the system with time can be inferred.

For n-dimensional motion, the position and velocity components are co-ordinates in a 2n-dimensional *phase space*. When $n = 1$, for example, the phase space is a plane, with mutually perpendicular axes x and v. Initially the particle can be described by a point in the phase plane given by (x_0, v_0). As time passes the point describing the particle moves along a particular path in the phase plane. For different conditions, the motion will be along different phase paths. No two phase paths can cross, since doing so would mean that a particle could follow two different motions for a given set of initial conditions, in violation of the fact that the motion is uniquely specified by the equation of motion and the initial conditions. The collection of all possible phase paths is known as the *phase portrait* of the dynamic system.

For one-dimensional motion under a conservative force, mechanical energy is conserved:

$$\tfrac{1}{2}mv^2 + U(x) = E \,,$$

where E is constant. Thus

$$v^2 = \frac{2}{m}[E - U(x)] . \tag{4.11}$$

From this equation we see that the phase portrait of a particle moving in one-dimension under the influence of a conservative force is symmetric about the x-axis.

For the simple harmonic oscillator Eq. 4.11 becomes

$$v^2 = \frac{2E}{m} - \frac{k}{m}x^2 .$$

This is the equation for an ellipse in phase space, so that the phase portrait consists of concentric ellipses.

One method for using Maple to create phase plots in one-dimension is to create position and velocity as explicit functions of the initial conditions, as well as of time. Continuing the earlier Maple dialogue, unassign the initial position and velocity, and define new functions X(x0,v0,t) and V(x0,v0,t) as follows.

```
> x0 := 'x0':      v0 := 'v0':
> X := unapply(x(t), x0, v0, t):
> V := unapply(diff(x(t),t), x0, v0, t):
```

Assuming that omega0 has been set to 1, as discussed earlier, we can generate phase paths for the simple harmonic oscillator by plotting V vs. X for various initial conditions. Since the motion is periodic, we need vary t over only one period to get complete paths. So, for example, the Maple commands

```
> omega0 := 1:      # if not done previously
> plot({[X(1,0,t), V(1,0,t), t=0..2*Pi], [X(2,0,t), V(2,0,t), t=0..2*Pi],
>       [X(3,0,t), V(3,0,t), t=0..2*Pi]});
```

generate the three phase paths shown in Fig. 4.2. The fact that the phase paths close upon themselves is an indication that the motion is periodic. This was known to be the case for the problem, and was used when limiting the range in time to $0 \le t \le 2\pi$. For more complex motion, however, the presence of closed paths is a highly useful indicator for periodic motion.

4.3   Damped Harmonic Motion

4.3.1   Equation of Motion and Its Solutions

If a frictional damping force proportional to the velocity, $\mathbf{f} = -b\mathbf{v}$, is added to the spring force of the harmonic oscillator and terms are rearranged, the

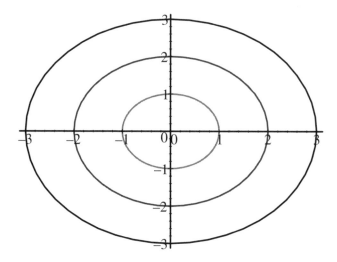

FIGURE 4.2. Phase plot for the simple harmonic oscillator.

x-direction equation of motion can be written

$$\ddot{x}(t) + 2\beta\dot{x}(t) + \omega_0^2 x(t) = 0 \,, \tag{4.12}$$

where

$$\beta = \frac{b}{2m}$$

is a damping constant having units of inverse time, and ω_0 was defined in Sec. 4.1.

Equation 4.12 is entered into Maple and solved in the manner presented in Sec. 4.2. We check and assign the solution in the usual way. The sequence is

```
> eq := diff(x(t),t,t) + 2*beta*diff(x(t),t) + omega0^2*x(t) = 0:
> sol := dsolve({eq, x(0)=x0, D(x)(0)=v0}, x(t)):
> simplify(subs(sol,eq));
```

$$0 = 0$$

```
> assign(sol):
```

The solution for x(t) consists of terms with exponential factors of the form

$$e^{\left(-\beta\pm\sqrt{\beta^2-\omega_0^2}\right)t} \,.$$

There are four possible classes of solution, depending on the relative sizes of β and ω_0. We looked at the $\beta = 0$ (undamped) case in Sec. 4.2. The other three cases, underdamped ($\beta < \omega_0$), overdamped ($\beta > \omega_0$), and critically damped ($\beta = \omega_0$) are each to be examined in turn.

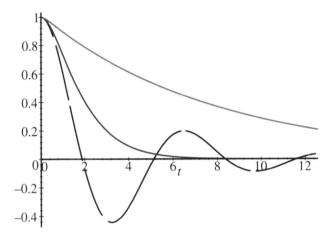

FIGURE 4.3. Position of the mass in damped harmonic motion for three values of the damping constant.

However, before studying the three damping cases individually, let us look at the behavior of the solution for some specific values of the parameters. We choose $v_0 = 0$ and $x_0 = 1$ for the initial values, which is equivalent to starting the oscillator from rest and choosing a length scale equal to the initial displacement. As before, let $\omega_0 = 1$, amounting to a change in the time scale. The only free parameter that remains is the damping constant. We plot solutions for $\beta = 0.25, 1, 4$, which fall into the underdamped, critically damped, and overdamped regions, respectively. The Maple commands

```
> x0 := 1:     v0 := 0:     omega0 := 1:
> plot({limit(x(t),beta=.25), limit(x(t),beta=1),
>      limit(x(t),beta=4)}, t=0..4*Pi);
```

result in the plot shown in Fig. 4.3. The gaps in the oscillatory curve ($\beta = .25$) in this figure are due to small imaginary parts arising from round-off errors in the floating-point calculations, as noted in Chapter 1. In this case using 1/4 rather than 0.25 solves the problem, but this will not be true in general since the plot variable t has floating-point values.

The three cases appear distinctly different for this set of parameters. In particular, the underdamped case is oscillatory, but with decreasing amplitude. The overdamped and critically damped cases monotonically decrease to zero. Different initial conditions, of course, show somewhat different behavior, but the oscillatory character of the underdamped case is maintained, as is the non-oscillatory character of the other two cases.

In order to examine the three cases for more general conditions, unassign x_0, v_0, and ω_0.

> x0 := 'x0':   v0 := 'v0':   omega0 := 'omega0':

Underdamped Case ($\beta < \omega_0$)

The underdamped case is most similar to the undamped, simple harmonic motion described in Sec. 4.2, and in fact, goes over to it in the $\beta \to 0$ limit. For this case, the arguments of the exponential factors are complex, with a negative real part, so that $x(t) \to 0$ as $t \to \infty$. To examine the solution with Maple, use the fact that $\beta < \omega_0$ to make the substitution $-\omega_1^2 = \beta^2 - \omega_0^2$. Since β, ω_0, and ω_1 are all positive, this is equivalent to the replacement $\omega_0 = \sqrt{\beta^2 + \omega_1^2}$. Since we assigned x(t) as described earlier, the following Maple commands produce a relatively simple form for x(t):

> assume(omega1>0):   assume(beta>0):
> subs(omega0=sqrt(beta^2+omega1^2), x(t)):
> simplify(evalc("));

$$\frac{(\sin(\omega 1\, t)\, x0\, \beta + x0\, \omega 1\, \cos(\omega 1\, t) + \sin(\omega 1\, t)\, v0)\, e^{(-\beta t)}}{\omega 1}$$

The evalc call causes Maple to give the results in terms of sin and cos functions rather than exponentials with complex arguments. (The ˜ symbols that Maple uses to indicate assumed variables have been suppressed for readability.)

Collecting terms,

> collect(", sin(omega1*t));

and rearranging slightly yields

$$x(t) = \frac{(\beta x_0 + v_0)}{\omega_1} e^{-\beta t} \sin \omega_1 t + x_0\, e^{-\beta t} \cos \omega_1 t . \qquad (4.13)$$

The separate sin and cos terms can be combined into a single trigonometric function with a phase angle if desired. In any case, it is clear that the motion is sinusoidally oscillating, with exponentially decreasing amplitude. The time constant for the exponential decrease is $1/\beta$.

Overdamped Case ($\beta > \omega_0$)

A simplified expression for the overdamped case can be obtained from the general solution assigned to x(t) by substituting $\alpha = \sqrt{\beta^2 - \omega_0^2}$; α is positive. This substitution can be made in the expression for x(t) and simplified with the sequence

```
> assume(alpha>0):
> subs(omega0=sqrt(beta^2 - alpha^2), x(t)):
> simplify(");
```

$$-\tfrac{1}{2}\left(-\beta\, e^{((-\beta+\alpha)t)}x0 + \beta\, e^{(-(\beta+\alpha)t)}x0 - e^{((-\beta+\alpha)t)}x0\,\alpha \right.$$
$$\left. + e^{(-(\beta+\alpha)t)}v0 - e^{((-\beta+\alpha)t)}v0 - e^{(-(\beta+\alpha)t)}x0\,\alpha\right)/\alpha$$

```
> collect(", [exp(-(beta-alpha)*t), exp(-(beta+alpha)*t)], factor);
```

The resulting expression for x(t) becomes, with minor manual simplification,

$$x(t) = \frac{[(\beta+\alpha)x_0 + v_0]}{2\alpha}e^{-(\beta-\alpha)t} - \frac{[(\beta-\alpha)x_0 + v_0]}{2\alpha}e^{-(\beta+\alpha)t} . \qquad (4.14)$$

Note that since $\beta > \alpha$, both terms are exponentially decreasing with time, as might have been expected from Fig. 4.3. The first term decreases more slowly, and so dominates the second for $t \gg 1/(\beta+\alpha)$. It also decreases more slowly than the underdamped case, except for initial conditions for which the first term of Eq. 4.14 vanishes, i.e., when $v_0 = -(\beta + \alpha)x_0$.

Critically Damped Case ($\beta = \omega_0$)

If the damping constant is the same as the natural angular frequency, we get the critically damped case. The solution in this case can be obtained from the general expression for x(t) by taking the limit,

```
> limit(x(t), omega0=beta);
```

The result is

$$x(t) = (x_0(1 + \beta t) + v_0 t)\, e^{-\beta t} . \qquad (4.15)$$

This solution indicates that for x_0 and v_0 positive, the position decreases to zero with increasing time. If $\beta x_0 + v_0$ has the opposite sign from x_0, $x(t)$ changes sign once and approaches zero from the other side of the t-axis as t gets larger.

4.3.2   Further Examination of the Underdamped Case

As we have seen, the major difference between the undamped harmonic oscillator and the underdamped oscillator is that the amplitude of oscillation steadily decreases for the latter case. This means that mechanical energy is not conserved, due, of course, to the frictional damping force. This is easily demonstrated with Maple by defining kinetic and potential energies and graphing each, along with their sum, as a function of time. For the conditions $x_0 = 1$, $v_0 = 0$, $\omega_0 = 1$, and $\beta = \frac{1}{10}$ in appropriate units, the

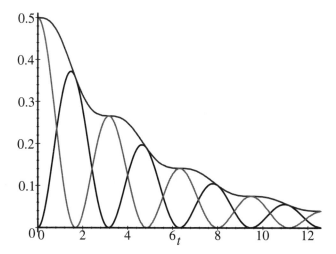

FIGURE 4.4. Kinetic, potential, and total energy as a functions of time for an under-damped harmonic oscillator.

result is shown in Fig. 4.4. In the figure, we see the energy alternating between potential and kinetic forms, with the sum, the mechanical energy, decreasing monotonically, though certainly not at a constant rate.

If you examine the figure, you will find that the mechanical energy is almost constant at points in time where the *potential* energy is near a local maximum, and is decreasing most rapidly in regions where the *kinetic* energy is near a local maximum. Why?

As a final comparison with the undamped case, we examine the phase plot of the underdamped harmonic oscillator. The result is shown in Fig. 4.5 using the same parameter values as for Fig. 4.2, with the addition of $\beta = \frac{1}{10}$. The time range for these phase paths is $0 \le t \le 4\pi$, which corresponds to two periods of oscillation of a similar undamped oscillator. In comparison with the phase plot for the undamped oscillator (Fig. 4.2), the obvious difference is the inward spiralling of the phase paths as energy is dissipated by the frictional force. The origin of the phase plot, which is the point where the mass is motionless and the spring is neither stretched nor compressed, is an *attractor* of this dynamic system.

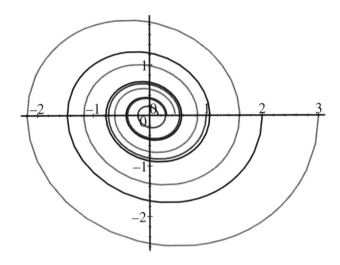

FIGURE 4.5. Phase plot of the damped harmonic oscillator.

4.4 Sinusoidally-Driven Harmonic Motion

4.4.1 *Solution of the Equation of Motion*

Consider now the case of a damped harmonic oscillator which is subjected to a sinusoidally-varying external force, $F_{ext} = F_0 \sin \omega t$. Using the same definitions as in Section 4.3.1, the equation of motion becomes

$$\ddot{x}(t) + 2\beta \dot{x}(t) + \omega_0^2 x(t) = \frac{F_0}{m} \sin \omega t \; . \tag{4.16}$$

This is a non-homogeneous, linear, second-order differential equation. The usual way to solve it is to add to the general solution of the corresponding homogeneous equation any particular solution to Eq. 4.16. A particular solution consisting of a combination of $\sin \omega t$ and $\cos \omega t$ terms can be found with a bit of algebra.

With Maple it is unnecessary to consider the homogeneous and particular solutions separately. We can enter Eq. 4.16 and use dsolve to solve it just as before. For variety, we do not specify initial conditions, and so get the general solution to the equation. The commands to solve and check the solution are

```
> eq := diff(x(t), t, t) + 2*beta*diff(x(t), t)
>       + omega0^2*x(t) = Fo/m*sin(omega*t):
> sol := dsolve(eq, x(t)):
> simplify(subs(sol,eq));
```

$$\frac{F_o \sin(\omega t)}{m} = \frac{F_o \sin(\omega t)}{m}$$

The resulting general solution for x(t) is somewhat involved. We can see its structure more clearly by collecting terms that depend on the driving force. We assign the result to x(t).

> assign(collect(sol, Fo, factor)):   x(t);

$$x(t) := \frac{1}{2} \frac{I\left(\omega0^2 - \omega0^2 e^{(2I\omega t)} + \omega^2 e^{(2I\omega t)} + 2Ie^{(2I\omega t)}\omega\beta + 2I\omega\beta - \omega^2\right)F_o}{me^{(I\omega t)}\left(4\omega^2\beta^2 - 2\omega^2\omega0^2 + \omega0^4 + \omega^4\right)}$$

$$+ \_C1\, e^{\left(\left(-\beta + \sqrt{(\beta-\omega0)(\beta+\omega0)}\right)t\right)} + \_C2\, e^{\left(-\left(\beta + \sqrt{(\beta-\omega0)(\beta+\omega0)}\right)t\right)}$$

Note that the terms that do not contain F_0 involve arbitrary integration constants, which Maple labels _C1 and _C2 (assuming this is a new Maple session). They depend on the natural angular frequency ω_0, but not on the driving angular frequency ω. They are the terms that form the general solution of the homogeneous equation. (See Section 4.3.1.) The term that is proportional to F_0 makes up a particular solution to the inhomogeneous Eq. 4.16. Although Maple has expressed it in terms of complex exponentials, it is in fact real. We can verify this as follows:

> convert(coeff(x(t), Fo), trig):
> simplify(expand(")):

$$-\frac{-\sin(\omega t)\,\omega0^2 + \omega^2\sin(\omega t) + 2\omega\beta\cos(\omega t)}{m\left(\omega^4 + 4\omega^2\beta^2 - 2\omega^2\omega0^2 + \omega0^4\right)}$$

Further examination of the solution reveals that the terms with arbitrary constants contain exponential factors that decrease with increasing time since $\beta > 0$, whereas the F_0 terms vary sinusoidally with constant amplitude. Because of this behavior, the exponentially decreasing terms are referred to as the *transient* part of the solution, or simply, transients. By contrast, the terms involving F_0 are referred to as the *steady-state* part of the solution, or simply, the steady-state solution. Note that the initial conditions for the motion provide values for the constants _C1 and _C2, but do not change the steady-state part. Thus, after a long time the motion is approximately the same, regardless of the initial conditions.

If we need the solution for particular initial conditions, we can solve for the appropriate values of _C1 and _C2. For example, suppose the initial position and velocity are x_0 and v_0. The appropriate values for _C1 and _C2 are found and assigned with the Maple commands

> solve(subs(t=0, {x(t)=x0, diff(x(t),t)=v0}), {_C1,_C2}):
> assign(simplify(")):

We can see the effect on x(t) by executing the command eval(x(t)). The eval is necessary because Maple does not fully evaluate assigned solutions to differential equations unless specifically requested to do so. (This is probably a bug. Curiously, if we assign x(t) to a new variable, and ask for the value of the new variable, the result *is* fully evaluated, with _C1 and _C2 replaced by their assigned values. The command collect(x(t), Fo) will also cause the constants to be replaced.)

An alternative to solving for _C1 and _C2 is to use dsolve to solve Eq. 4.16 together with initial conditions.

```
> x(t) := 'x(t)':
> dsolve({eq, x(0)=x0, D(x)(0)=v0}, x(t)):
> assign(collect(", Fo, simplify)):
```

4.4.2   Energy, Power, and Resonance

We can gain a good deal of insight from examining the driven harmonic oscillator from an energy point of view. To this end, we use the value given to x(t) by the assign command at the end of the previous section to first define the velocity of the oscillator, and then the kinetic and potential energies.

```
> v := diff(x(t), t):
> K := 1/2*m*v^2:
> U := 1/2*m*(omega0*x(t))^2:
```

In order to plot results, we assign numerical values to the parameters; for example,

```
> m := 1:     omega0 := 1:    Fo := 1:     beta := 1/10:
> x0 := 0:    v0 := 0:
```

The assignments on m and omega0 are equivalent to changes of scale, and so do not limit qualitative or quantitative analysis. The other assignments do limit analysis to special cases, although the resulting behavior should qualitatively apply to other conditions as well. Note that these values for ω_0 and β correspond to a lightly damped oscillator.

Let us first look at the mechanical energy of the oscillator as a function of time for three values of ω. This is easily done by defining the energy to be a function of ω and t.

```
> En := unapply(Re(K+U), omega, t):
```

The exact K+U is of course real. However, without the Re in the definition, floating-point arguments to En, such as 0.5 or 0.9 below, or the values for t selected by Maple in the plotting process, can cause En to return a small imaginary component due to floating-point roundoff errors. plot cannot successfully handle complex arguments.

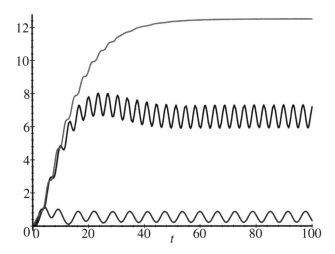

FIGURE 4.6. Energy of the sinusoidally-driven harmonic oscillator as a function of time for three values of the driving frequency.

Plotting the energy as a function of time for $\omega = 0.5, 0.9$, and 1 (units of ω_0),

```
> plot({En(0.5,t), En(0.9,t), En(1,t)}, t=0..10/beta);
```

yields Fig. 4.6. We have previously seen that the energy of the undriven, undamped oscillator is constant, and that of the undriven, damped oscillator decreases in time. Thus, it should be obvious that the initial increase in energy in this case comes from the driving force. Note also that the energy settles into regular, periodic behavior at longer times. This is the steady-state region. Since the transient part of the underdamped solution consists of sinusoidal terms multiplied by $e^{-\beta t}$ factors, the steady-state terms of the position and velocity (and thus energy) should be dominant for $t \gg 1/\beta$.

Notice that of the three ω-values shown, the energy absorption is greatest at $\omega = \omega_0$. This is true not only for the three values chosen for Fig. 4.6, but for all ω-values. A three-dimensional plot of the energy as a function of ω and t better illustrates this selective absorption near $\omega = \omega_0$. Figure 4.7 was created with the Maple command

```
> plot3d(En(omega,t), t=0..30, omega=0..2, grid=[60,40],
>       axes=frame, orientation=[-60,60], view=0..12);
```

The axes and orientation arguments are rather easily set interactively, and so may be omitted.

The preferential absorption near $\omega = \omega_0$ seen in Fig. 4.7 is a manifestation of the phenomenon of *resonance*: a driven oscillator is most strongly affected

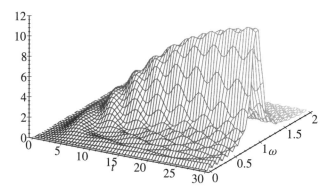

FIGURE 4.7. Energy of the sinusoidally-driven harmonic oscillator as a function of the driving frequency and time.

by a periodic driving force that has the same or nearly the same frequency as the natural frequency of the oscillator. This effect can be illustrated more clearly by examining the power absorbed by the oscillator from the driving force, averaged over one period of the driving force:

$$ P_{avg} = \frac{1}{T} \int_{t_0}^{t_0+T} v(t) F_0 \sin \omega t \, dt \,, $$

where $v(t)$ is the velocity of the oscillator, $T = 2\pi/\omega$ is the oscillation period for the driving force, and t_0 is the starting time for the period over which the average is taken.

It is instructive to compare the power absorbed for different values of damping, so we unassign β and define the average power absorbed as a function of β and ω. Two cases with different starting times, $t_0 = 0$ and $t_0 = 10/\beta$, are considered. By $t = 10/\beta$ the transients have effectively died out, whereas at $t = 0$ they are very much present.

```
> beta := 'beta':
> T := 2*Pi/omega:
> int(v*Fo*sin(omega*t), t=t0..t0+T)/T:
> AvgPower := unapply(", t0, beta, omega):
> P0 := (beta,omega) -> Re(AvgPower(0, beta, omega)):
> Pss := (beta,omega) -> Re(AvgPower(10/beta, beta, omega)):
```

P0 gives the average power absorbed over a period starting at $t = 0$, when transients are present; Pss gives the average power absorbed over a period starting at $t = 10/\beta$, which is near steady state. The Re function is applied in each case to extract the real part since a small imaginary part can arise when AvgPower is evaluated at floating-point values of β or ω. If now the average power is plotted as a function of ω, the effect of a driving force

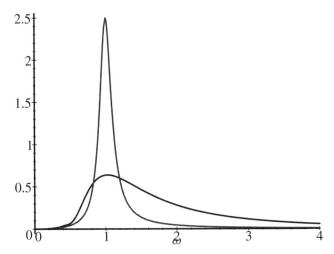

FIGURE 4.8. Average power absorbed by the sinusoidally-driven harmonic oscillator as a function of driving frequency. The two curves represent different initial times for the average.

whose frequency is at or near resonance can be clearly seen.

> plot({P0(.1,omega), Pss(.1,omega)}, omega=0..4);

The result is shown in Fig. 4.8, for the case $\beta = \frac{1}{10}$, which corresponds to weak damping. Notice that when transients are still present, significant power absorption can occur for a broad range of driving frequencies. As steady state is approached, however, power absorption is much larger for driving frequencies near the natural frequency.

Finally, Fig. 4.9 shows a comparison of near-steady-state power absorption for three different values of β. The plot is that obtained from

> plot({Pss(.1,omega), Pss(.25,omega), Pss(.5,omega)}, omega=0..4);

Notice again the strong power absorption near resonance ($\omega = \omega_0 = 1$), and the large increase in resonant absorption as damping is decreased.

Keeping only the steady-state part of the solution for x(t) by setting _C1 = _C2 = 0 in the general solution, use Maple to show that the average steady-state power absorption at resonance is $F_0^2/(4m\beta)$. Give a physical explanation for why the res-

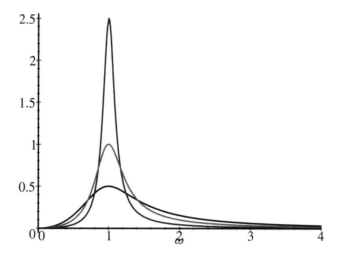

FIGURE 4.9. Average steady-state power absorption of the sinusoidally-driven harmonic oscillator as a function of driving frequency. The three curves represent different values of the damping constant.

onant steady-state power absorption increases with decreasing damping.

4.5   Impulse-Driven Harmonic Oscillator

As another example of a driven oscillator, consider a damped harmonic oscillator driven by a series of brief impulses. To model such a driving force we use a series of Dirac delta functions, each multiplied by a constant p_0 with units of momentum:

$$F(t) = \sum_{j=1}^{n} p_0 \delta(t - j\tau) .$$

Because of the integral properties of the delta function, each impulse instantaneously imparts an additional momentum of p_0 to the oscillator at times $\tau, 2\tau, 3\tau, ..., n\tau$. The frequency of the impulses is $1/\tau$.

The equation of motion can be written

$$\ddot{x}(t) + 2\beta\dot{x}(t) + (\omega_1^2 + \beta^2)x(t) = \frac{F(t)}{m} . \qquad (4.17)$$

Anticipating the underdamped solution for the homogeneous equation, we replace ω_0^2 by $\omega_1^2 + \beta^2$. This nudges Maple into a more convenient form for the solution, without the necessity of the manipulations done in the previous treatment of the underdamped case.

The Dirac delta function is included in Maple's library, and with the optional laplace argument to dsolve Maple can solve Eq. 4.17 for a specific value for n. The approach here is to use Maple to solve the equation for $n = 1$, identify the homogeneous and particular parts of the solution, and then use the superposition theorem to write the general solution for the multiple-impulse case. For concreteness, consider initial conditions of $x(0) = x_0$ and $\dot{x}(0) = 0$. The Maple commands given below solve the equation and write the solution as a Maple function of n, τ, and t.

```
> eq := diff(x(t),t,t) + 2*beta*diff(x(t),t)
>       + (omega1^2+beta^2)*x(t) = Fdrive/m:
> Fdrive := p0*Dirac(t-t0):
> assume(t0>0):
> sol := dsolve({eq, x(0)=x0, D(x)(0)=0}, x(t), laplace);
```

$$sol := x(t) = \frac{x0\,\beta\,e^{(-\beta t)}\sin(\omega 1\,t)}{\omega 1} + x0\,e^{(-\beta t)}\cos(\omega 1\,t)$$
$$+ \frac{p0\,Heaviside(t-t0\tilde{})e^{(-\beta(t-t0\tilde{}))}\sin(\omega 1(t-t0\tilde{}))}{m\,\omega 1}$$

```
> xh := rhs(subs(p0=0, sol));              # homogeneous solution
```

$$xh := \frac{x0\,\beta\,e^{(-\beta t)}\sin(\omega 1\,t)}{\omega 1} + x0\,e^{(-\beta t)}\cos(\omega 1\,t)$$

```
> subs({x0=0, t0=j*tau}, sol):
> xp := Sum(rhs("), j=1..n);               # particular solution
```

$$xp := \sum_{j=1}^{n} \frac{p0\,Heaviside(t-j\tau)e^{(-\beta(t-j\tau))}\sin(\omega 1(t-j\tau))}{m\,\omega 1}$$

```
> X := unapply(xh+xp, n, tau, t):
```

The Heaviside function that appears in the particular part of the general solution is a step function defined by

$$Heaviside(x) = \begin{cases} 0 & x < 0 \\ 1 & x \geq 0 \end{cases}.$$

Let us look at the effect of the driving force on the position of the mass. To this end, we make assignments to the parameters of the problem in order to plot $x(t)$.

```
> x0 := 1:    p0 := 1:    beta := 1/10:    omeg0 := 1:
> omega1 := sqrt(omega0^2-beta^2):    T:=2*Pi/omega1:
```

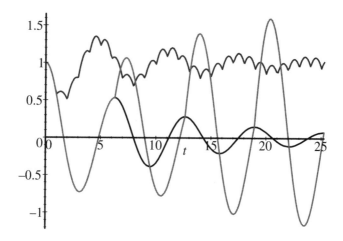

FIGURE 4.10. Position of the mass of the impulse-driven harmonic oscillator as a function of time for three values of the impulse frequency.

Figure 4.10 shows the position of the mass as a function of time for the damped, undriven oscillator, and the impulse-driven oscillator with $\tau = 1/\omega_0$ and $\tau = T = 2\pi/\omega_1$. The latter value for τ is the period of the oscillatory part of the undriven damped oscillator solution, and is approximately the time between positive peaks of the undriven oscillator. For each of the driven oscillator cases shown in the figure the number of impulses n is chosen so that the impulses continue at regular intervals over the entire time range (4T). The figure is obtained with the Maple command

```
> plot({X(0,0,t), X(trunc(4*T),1,t), X(4,T,t)}, t=0..4*T);
```

Compare the $\tau = T$, impulse-driven solution with the undriven solution. As expected, the two solutions coincide until the second positive peak, $t = T$, when the first impulse arrives. The sign and magnitude of the impulse is such that it gives the mass sufficient additional velocity to carry it to a peak larger than the starting position. Subsequent impulses arrive at approximately the times where the undriven oscillator peaks in the positive direction. Each of these impulses thus adds energy to the oscillator. If the energy added is greater than that lost due to damping since the last impulse, the amplitude of oscillation will increase with each impulse, as it does here. This will be reflected in the plot of the energy to be shown shortly.

The remaining curve illustrates the effect of an oscillator which is driven by impulses that come more frequently than the oscillation time of the undriven oscillator. In particular, for this case there are about six impulses between each positive peak. Consequently, the impulses sometimes add energy to the oscillator, and sometimes remove energy, causing the position of the mass to oscillate weakly about some average displacement.

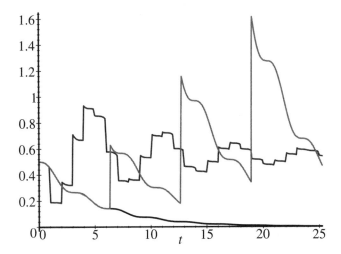

FIGURE 4.11. Energy of the impulse-driven harmonic oscillator as a function of time for three values of the driving frequency.

These conclusions are amplified by an energy plot. We define the velocity and mechanical energy functions

```
> V := unapply(diff(X(n,tau,t),t), n, tau, t):
> m := 1:    k := 1:
> En := (n,tau,t) -> 1/2*m*V(n,tau,t)^2 + 1/2*k*X(n,tau,t)^2:
```

to aid us in plotting the energy for varying values of the impulse time and a given number of impulses.

Figure 4.11 is a plot of the energy of the damped, undriven oscillator and two impulse-driven cases. The parameters used are the same as for the previous figure. The Maple command to produce this figure is

```
> plot({En(0,0,t), En(trunc(4*T),1,t), En(4,evalf(T),t)}, t=0..4*T);
```

The evalf(T) is required in this statement because Maple does not recognize that a Dirac function argument which depends on the irrational number π is non-zero, and that thus it should return a value of zero for the function. By forcing floating-point evaluation, the argument is recognized as non-zero, and the appropriate value for the Dirac function is returned.

The energy of the undriven case drops monotonically to zero, with the plateaus and rapid drops characteristic of the underdamped case, as seen in Sec. 4.3. The oscillator which is driven by impulses at times that are integer multiples of $\tau = 1$, sometimes gains, and sometimes loses energy when the impulse arrives. Consequently, its average energy remains approximately constant. By contrast, the oscillator driven in such a way that the impulses

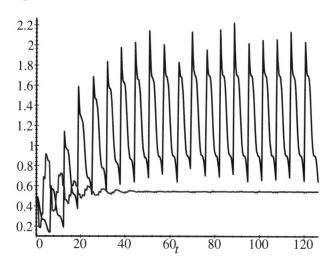

FIGURE 4.12. Energy of the impulse-driven harmonic oscillator as a function of time for three values of the driving frequency.

arrive at or near the peak positive value of the displacement always receives a boost in energy. Although the energy decreases because of the damping between impulses, the increase is greater than the amount lost, causing the average energy of the oscillator to continue to increase.

As in the sinusoidally-driven case, the increase in oscillation amplitude and energy of the damped, resonantly-driven oscillator does not continue indefinitely. Since the dissipative force is proportional to the velocity of the mass, the power dissipated is proportional to the kinetic energy at each instant. Thus, as time progresses, each impulse adds a constant amount of energy to the oscillator, but the energy lost between impulses grows larger. At very long times the driven oscillator approaches steady state. This is true for any driving frequency, but the steady-state energy will be much larger for oscillators driven at or near resonance. Figure 4.12 shows the plot of the energies for the same driving frequencies as for Fig. 4.11, but over the extended range $0 \leq t \leq 20T$. The command is

```
> plot({En(trunc(20*T),1,t), En(20,evalf(T),t)}, t=0..20*T);
```

There are two things to note about this plot. First, it takes quite a few minutes to compute. Second, the details of the resonantly-driven curve are very dependent upon the points sampled in the plotting because of the discontinuities in the energy. However, the curve is qualitatively correct, as can be seen by repeating the plot with more points using the optional numpoints argument.

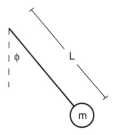

FIGURE 4.13. Schematic of the simple pendulum.

4.6 Approximate Simple Harmonic Motion

4.6.1 The Simple Pendulum

Consider the motion of the simple pendulum: a mass attached to one end of a massless rigid rod. The other end of the rod is pivoted so that the mass may swing in a vertical plane. This is a problem best suited to polar coordinates, with the origin at the site of the pivot. In this system $\rho = L$, the (constant) length of the rod. The angular displacement ϕ is measured from the vertical, as seen in Fig. 4.13.

Two forces act on the mass: gravity and the tension in the rod. There is no radial motion. The tangential component of the acceleration, as given in Eq. 3.8, leads to the equation of motion

$$-mg \sin \phi = mL\ddot{\phi} \ .$$

Slight rearrangement of this equation leads to the following equation of motion for the simple pendulum:

$$\ddot{\phi} + \frac{g}{L} \sin \phi = 0 \ . \tag{4.18}$$

Notice that the motion depends on the single parameter g/L rather than g and L separately. We will use this fact later when we integrate Eq. 4.18 numerically.

Equation 4.18 cannot be solved in terms of simple functions. However, if the maximum angle of oscillation is small (i.e., $\phi \ll 1$), we can approximate $\sin \phi$ by the first term in its Taylor series expansion, yielding the following approximation to the equation of motion:

$$\ddot{\phi} + \frac{g}{L} \phi = 0 \ .$$

This is of the same mathematical form as the simple harmonic oscillator equation, Eq. 4.3, with $\omega_0 = \sqrt{g/L}$. Thus, the approximate solution may be written as

$$\phi(t) = \phi(0) \cos\left(\sqrt{\tfrac{g}{L}}t\right) + \dot{\phi}(0)\sqrt{\tfrac{L}{g}} \sin\left(\sqrt{\tfrac{g}{L}}t\right) \ .$$

In this lowest-order approximation the period of oscillation is $T = 2\pi\sqrt{L/g}$, which is independent of the amplitude of the oscillation.

However, it has been known since the time of Galileo that the period of simple pendulum motion depends on the amplitude of the swing. Corrections to this lowest-order expression may be found by considering conservation of energy for the simple pendulum. Energy is conserved in this problem since gravity is a conservative force, and the tension in the rod does no work on the mass. If we take the zero level for the gravitational potential energy to be at the bottom of the pendulum's swing,

$$U = mgL(1 - \cos\phi) = 2mgL\sin^2(\tfrac{1}{2}\phi) ,$$

conservation of energy gives

$$\tfrac{1}{2}mL^2\dot{\phi}^2 + 2mgL\sin^2(\tfrac{1}{2}\phi) = E .$$

Solving this equation for $\dot{\phi}$ yields

$$\dot{\phi} = \pm\sqrt{\frac{2E}{mL^2} - \frac{4g}{L}\sin^2(\tfrac{1}{2}\phi)} . \qquad (4.19)$$

If the motion is oscillatory, so that it remains in the interval $-\pi < \phi < \pi$, and ϕ_{max} is the positive angle at which the pendulum comes to a rest (i.e., the amplitude of the oscillation), the total energy E can be expressed as

$$E = 2mgL\sin^2(\tfrac{1}{2}\phi_{max}) .$$

Substituting this into Eq. 4.19 yields

$$\frac{d\phi}{dt} = \pm 2\sqrt{\frac{g}{L}\left[\sin^2(\tfrac{1}{2}\phi_{max}) - \sin^2(\tfrac{1}{2}\phi)\right]} .$$

Isolating the ϕ dependence on the left side and the t dependence on the right side, and integrating over time, we get the following equation for $\phi(t)$:

$$\int_{\phi_0}^{\phi(t)} \frac{d\phi}{\pm\sqrt{\sin^2(\tfrac{1}{2}\phi_{max}) - \sin^2(\tfrac{1}{2}\phi)}} = 2\sqrt{\frac{g}{L}}t , \qquad (4.20)$$

where $\phi_0 = \phi(0)$. The + sign applies to the counterclockwise part of the motion, while the - sign applies to the clockwise part.

Unfortunately, the integral in Eq. 4.20 cannot be evaluated analytically. It is an elliptic integral of the first kind, and can be put into standard form with the transformation that we will make shortly. However, Eq. 4.20 can be used to obtain corrections to the harmonic oscillator approximation for the period of the simple pendulum. To this end, we first take the limits of the integral on the left side of Eq. 4.20 to be $\phi_0 = 0$ and $\phi(t) = \phi_{max}$. The value of t appropriate to these limits is one-fourth of the period T. Furthermore, since $\phi_{max} > 0$, this is motion in the counterclockwise direction so that we choose the $+$ sign. Making these substitutions and solving for T, we find

$$T = 2\sqrt{\frac{L}{g}} \int_0^{\phi_{max}} \frac{d\phi}{\sqrt{\sin^2(\frac{1}{2}\phi_{max}) - \sin^2(\frac{1}{2}\phi)}} \, .$$

Changing variables from ϕ to χ with the transformation

$$\sin \chi = \frac{\sin\left(\frac{1}{2}\phi\right)}{\sin\left(\frac{1}{2}\phi_{max}\right)}$$

we obtain the following result for the period:

$$T = 4\sqrt{\frac{L}{g}} \int_0^{\frac{\pi}{2}} \frac{d\chi}{\sqrt{1 - \sin^2(\frac{1}{2}\phi_{max})\sin^2 \chi}} \, . \tag{4.21}$$

The integral is in the canonical form for the elliptic integral of the first kind[1], resulting in

$$T = \sqrt{\frac{L}{g}} F\left(\tfrac{1}{2}\pi \setminus \tfrac{1}{2}\phi_{max}\right) \, .$$

The elliptical integral $F(...\setminus...)$ is well-tabulated, so that the period can be found for any given value of ϕ_{max}. Alternatively, we can ask Maple to perform a numerical evaluation of the integral with evalf(Int(...)). However, here it is more revealing to expand the integrand in Eq. 4.21 in a Taylor series in ϕ_{max}, and integrate term-by-term. This is easily done with Maple; the commands are

```
> T := 4*sqrt(L/g)*int(1/sqrt(1 - sin(Phi/2)^2*sin(chi)^2), chi=0..Pi/2):
> To := 2*Pi*sqrt(L/g):
> series(T/To, Phi, 5);
```

The result to fourth order in ϕ_{max} is

$$T \simeq 2\pi\sqrt{\frac{L}{g}}\left(1 + \tfrac{1}{16}\phi_{max}^2 + \tfrac{11}{3072}\phi_{max}^4\right) \, .$$

As can be seen from this result, the period increases with increasing amplitudes, though even for ϕ_{max} as large as $\pi/4$ the correction is less than 4%. For values of $\phi_{max} > 1$ this truncated series becomes increasingly inadequate. Indeed, as $\phi_{max} \to \pi$, $T \to \infty$, as can be verified with Eq. 4.21. Physically this occurs because the tangential component of the gravitational force, and hence the acceleration, becomes negligibly small as ϕ approaches π. In fact,

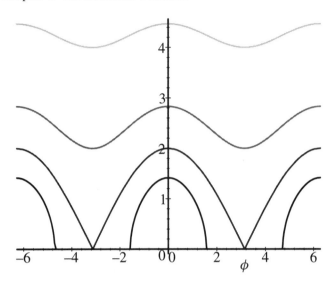

FIGURE 4.14. Phase plot (upper portion) for the simple pendulum.

$\phi = \pi$ is a point of unstable equilibrium so that if the mass comes to rest at that point it remains there indefinitely, unless disturbed by some other force.

The phase portrait of the simple pendulum is quite interesting and informative. It is easily obtained by using Eq. 4.19 to plot $\sqrt{L/g}\,\dot{\phi}$ vs. ϕ for several values of the energy. For $E < 2mgL$ the right side of Eq. 4.19 becomes imaginary for values of $\phi > \phi_{max}$. Only the real part is relevant, so we create the phase plot with Maple by first using Eq. 4.19 to define the function

```
> Phidot := (e,phi) -> Re(sqrt(2*e - 4*sin(phi/2)^2)):
```

This is a unitless form for $\dot{\phi}$, with the scaled energy $e = E/(mgL)$ and t in units of $\sqrt{L/g}$. If we now plot this scaled $\dot{\phi}$ for several values of e,

```
> plot({Phidot(1,phi), Phidot(2,phi), Phidot(4,phi), Phidot(10,phi)},
>       phi=-2*Pi..2*Pi);
```

we get the curves shown in Fig. 4.14. Note that in defining Phidot we have used only the positive root for $\dot{\phi}$, so Fig. 4.14 shows only the upper half of the complete phase portrait.

Mentally adding the lower part of the phase portrait, we see that there are two qualitatively different kinds of motion, corresponding to phase paths which close upon themselves (*i.e.*, oscillatory motion), and those that do not. The two regions are separated by the $E = 2mgL$ ($e = 2$) case. For energies less than this amount the pendulum oscillates between the two turning points $\phi = \pm\phi_{max}$, where the amplitude ϕ_{max} is determined by $E = 2mgL\sin^2(\frac{1}{2}\phi_{max})$. For small amplitude oscillations the phase paths are nearly ellipses, as expected for simple harmonic motion; but as ϕ_{max} approaches π, the paths become more and more distorted as the non-linear terms in the force begin to come into play. $E = 2mgL$ ($\phi_{max} = \pi$) yields the curve of Fig. 4.14 which separates the phase plane into regions of oscillatory and non-oscillatory motion. This curve is not a true phase path, since for this value of E the points at $\phi = \pm\pi$ are points of unstable equilibrium, where the pendulum comes to a rest.

For $E > 2mgL$, the pendulum has sufficient energy to move completely in a circle. Under these circumstances ϕ will increase or decrease monotonically, depending on the sign of $\dot{\phi}$. As seen in the figure, the angular velocity will oscillate, reaching a maximum as the mass passes through its lowest point ($\phi = 0, \pm2\pi, \pm4\pi$, etc.), where the potential energy is minimum. As E gets larger, the kinetic energy begins to dominate the potential energy. The oscillations in the angular velocity become less noticeable, and the phase paths become nearly straight.

4.6.2   Numerical Solution for the Simple Pendulum

Scaling the equation of motion for the simple pendulum, Eq. 4.18, so that time is in units of $\sqrt{L/g}$, yields the following scaled equation of motion:

$$\ddot{\phi} + \sin\phi = 0 . \tag{4.22}$$

As noted in the last section, this equation has no closed-form solution for $\phi(t)$. To partially overcome this, we solved the small-angle approximation to it, and examined a way to get higher-order corrections to the period of the motion. This section takes another look at the motion of the simple pendulum by numerically solving Eq. 4.22 and comparing with the results of the exact solution to the equation obtained by approximating $\sin\phi$ by ϕ (the simple harmonic oscillator approximation). For concreteness, we choose the initial angular coordinate $\phi(0) = \phi_0$, and the initial angular velocity $\dot{\phi}(0) = 0$. We then plot the comparison for two values of ϕ_0 ($\pi/2$ and $7\pi/8$ radians).

To solve with Maple, we first define the equation, and then use dsolve with the optional numeric argument.

```
> eq := diff(phi(t), t, t) + sin(phi(t)) = 0:
> sol1 := dsolve({eq, phi(0)=Pi/2, D(phi)(0)=0}, phi(t), numeric):
```

Recall from Chapter 1 that sol1 is now a function that when given a numerical value for its argument will return a list of the form [t=..., $\phi(t)$=..., $\partial\phi(t)/\partial t$=...]. To obtain ϕ as a function of time we extract the right-hand-side of the second element of sol1.

```
> Phi1 := s -> rhs(sol1(s)[2]):
```

The result of evaluating the function Phi1 at any value of its argument is the value of ϕ at that time value. For example, suppose we have defined sol1 and Phi1 as above and evaluate them at $t = 2.0$.

> sol1(2.0);

$$\left[t = 2.0, \ \phi(t) = -.2056395023696050, \ \frac{\partial}{\partial t}\phi(t) = -1.399235962023696 \right]$$

> Phi1(2.0);

$$-.2056395023696050$$

If we now define sol2 and Phi2 in similar ways, changing only the initial value of $\phi(0)$ in the dsolve call to $7\pi/8$,

> sol2 := dsolve({eq, phi(0)=7*Pi/8, D(phi)(0)=0}, phi(t), numeric):
> Phi2 := s -> rhs(sol2(s)[2]):

we get numerical solutions to Eq. 4.22 for two different amplitudes. They can be compared to the corresponding solution to the small-angle-approximation equation, $\phi_0 \cos(t)$. The comparison is most easily made by dividing each function by its value at $t = 0$ and plotting:

> plot('{cos(t), 2*Phi1(t)/Pi, 8*Phi2(t)/(7*Pi)}', t=0..4*Pi);

The single quotes are necessary to prevent Maple from attempting to evaluate Phi1(t) and Phi2(t) until a numerical value is available for t.

The results are shown in Fig. 4.15. It is easily seen that the primary result of the larger initial value for ϕ (since all curves are normalized to the same $t = 0$ value) is the increase in the period of the oscillation. There is some deviation of the curves from a simple cosine function, but it is difficult to see. The deviations get larger for larger initial angles, as expected based on the phase plot of Fig. 4.14.

The effect can be more clearly illustrated by comparing the 8*Phi2(t)/(7*Pi) plot with that of a cos function of the same period. First we determine the period of Phi2(t) by looking for the point where it reaches its second peak:

> dPhi2 := s -> rhs(sol2(s)[3]):
> T2 := fsolve('dPhi2(t) = 0', t=11..13);

$$T2 := 12.16080450$$

The comparison of the simple pendulum solution with that of a simple harmonic oscillator of the same period is then obtained from

> plot('{cos(2*Pi*t/T2), 8*Phi2(t)/(7*Pi)}', t=0..T2);

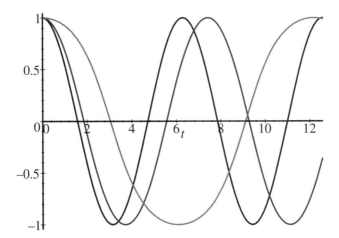

FIGURE 4.15. Angular position (relative to the maximum angular displacement) as a function of time for the simple pendulum. Two values for the maximum angular displacement are shown, along with the simple harmonic (small angle) approximation.

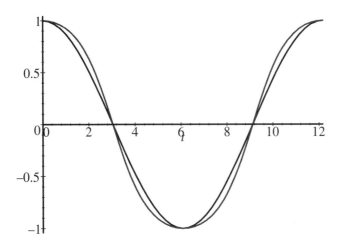

FIGURE 4.16. Detailed comparison of the angular position as a function of time between the simple pendulum and a harmonic oscillator of the same period.

which leads to Fig. 4.16. Clearly the simple pendulum spends somewhat more time near the extremes of its motion than a harmonic oscillator of the same period.

Consider the restoring torque on the simple pendulum as compared to a harmonic oscillator of the same period and explain physically the deviations shown in Fig. 4.16.

4.7 Problems

1. A mass m on a frictionless table is attached to a string. The other end of the string is passed through a hole in the table and tied to a second mass M, which hangs vertically. The mass on the table is given a velocity in the tangential direction of magnitude v_0.

   (a) Find the equations of motion for m and M.

   (b) Find the radius r_0 at which m will move in uniform circular motion, so that M remains stationary.

   (c) Assume that the mass m is started with the same tangential velocity, but moves at a radius $r(t) = r_0 + \Delta(t)$ which differs by a small amount from the equilibrium value. Substitute into the equations of motion to find the differential equation for $\Delta(t)$.

   (d) By keeping lowest order terms in $\Delta(t)$ show that the mass M undergoes approximate harmonic motion. Find the frequency of the oscillations.

2. A damped harmonic oscillator of mass m is subject to a driving force $F(t) = F_0 \sin^2 \omega t$.

   (a) Find the general solution for $x(t)$ from the equation of motion.

   (b) Obtain the steady state solution for $x(t)$.

   (c) Find the average power absorbed in steady state by the oscillator as a function of ω. Compute the average over one period (π/ω) of the driving force. You may find that you need to use convert(..., exp) to obtain an expression for the integrand that Maple can integrate.

   (d) Determine the resonant frequency of this driven oscillator by finding the frequency at which the average absorbed power is maximum.

   (e) Choose numerical values for the parameters and plot the average power as a function of ω.

3. Consider a mass m which is acted upon by Hooke's law forces in the x- and y-directions with force constants k_x and k_y respectively. Assume that the mass is moved to the point $(x_0, 0)$ and given an initial velocity of magnitude v_0 in the y-direction.

   (a) Find the solutions for $x(t)$ and $y(t)$.

(b) With an appropriate choice of units (by scaling of the equations of motion) you can take $m = 1$, $x_0 = 1$, and $\omega_x = \sqrt{k_x/m} = 1$. Assume that the initial velocity is also 1 in these units. Plot the trajectory of the mass over the range $0 \leq t \leq 10$ for several values of the ratio ω_y/ω_x. Make sure that for some of your choices the ratio is a simple rational number (e.g., 1, 2, 1/4, 5/3), and for some choices the ratio is irrational (e.g., π, $\sqrt{2}$, $1/\sqrt{3}$). Discuss the qualitative difference between the two types of trajectories.

4. A mass m is attached to a spring of spring constant k and whirled around in a horizontal plane at a constant angular speed ω. Neglect the effect of gravity.

(a) Find the equation of motion and discuss the three qualitatively different kinds of motion that result for the cases $\omega < \sqrt{k/m}$, $\omega = \sqrt{k/m}$, and $\omega > \sqrt{k/m}$.

(b) Solve the equation of motion for the initial conditions $r(0) = r_0$ and $\dot{r}(0) = v_0$. Verify the $\omega = \sqrt{k/m}$ limit for the solution.

(c) Let $k = 1$ and $m = 1$. (Note that this is just a change of scale.) Choose a set of initial conditions and plot $r(t)$ for several values of ω, making sure that each kind of motion is represented in your set of ω.

5. The deviations of simple pendulum motion from that of simple harmonic motion can be more dramatically seen by comparing the angular *velocity* of a simple pendulum with that of a harmonic oscillator of the same period and amplitude.

(a) Create such a plot, using the initial angular displacement $\phi(0) = \frac{7\pi}{8}$.

(b) At what part of the pendulum's motion are the differences greatest?

(c) Although the amplitude and period of the pendulum and simple harmonic oscillator are the same, their energies are not. Analytically compare the potential energies of the two systems at maximum displacement. Check that the result is in agreement with the numerical results for the kinetic energies at zero displacement.

5 Systems of Particles

5.1 The Two-Body Problem

Let us consider a system of two particles, masses m_1 and m_2, which interact with each other through a central force; that is, m_1 is acted upon by a force whose magnitude depends only upon the distance between the two particles, and whose center is at the position of m_2. The force on m_2 has the same form, with its center at m_1. The distance between the two particles is given by $r = |\mathbf{r}_1 - \mathbf{r}_2|$, where $\mathbf{r}_1$ and $\mathbf{r}_2$ are the positions of the two masses with respect to some inertial reference system. If we let $\hat{\mathbf{r}} = (\mathbf{r}_1 - \mathbf{r}_2)/r$ be the unit vector pointing from m_2 to m_1, the force acting on m_1 due to m_2 is of the form $\mathbf{F}_{12} = f(r)\hat{\mathbf{r}}$. Similarly, the force on m_2 due to m_1 is $\mathbf{F}_{21} = -f(r)\hat{\mathbf{r}}$. The equations of motion for the two masses are thus

$$f(r)\hat{\mathbf{r}} = m_1\ddot{\mathbf{r}}_1 \qquad (5.1)$$

and

$$-f(r)\hat{\mathbf{r}} = m_2\ddot{\mathbf{r}}_2 . \qquad (5.2)$$

Adding the corresponding sides of Eqs. 5.1 and 5.2 together, we find that the center of mass moves with constant velocity, as it should since there are no external forces:

$$M\ddot{\mathbf{R}} = 0 , \qquad (5.3)$$

where $M = m_1 + m_2$ is the total mass, and

$$\mathbf{R} = \frac{m_1\mathbf{r}_1 + m_2\mathbf{r}_2}{M}$$

is the position vector for the center of mass of the two particles.

Dividing each side of Eq. 5.1 by m_1 and each side of Eq. 5.2 by m_2, and subtracting the corresponding sides of the resulting equations yields

$$\left(\frac{1}{m_1} + \frac{1}{m_2}\right) f(r)\hat{\mathbf{r}} = \ddot{\mathbf{r}},$$

where $\mathbf{r} = \mathbf{r}_1 - \mathbf{r}_2$. This leads to the equation

$$f(r)\hat{\mathbf{r}} = \mu\ddot{\mathbf{r}}. \tag{5.4}$$

The quantity

$$\mu = \frac{m_1 m_2}{m_1 + m_2}$$

is the *reduced mass* of the two-particle system.

Notice what has occurred. We have transformed the two coupled equations of motion, Eqs. 5.1 and 5.2, into two uncoupled equations. Equation 5.3 describes the motion of the center of mass of the system, and Eq. 5.4 is equivalent to a single-particle, central-force equation of motion for a fictitious particle of mass μ. If we solve these two uncoupled equations for $\mathbf{R}$ and $\mathbf{r}$, we obtain the positions of the two masses with the coordinate transformations

$$\mathbf{r}_1 = \mathbf{R} + \frac{m_2}{m_1 + m_2}\mathbf{r}$$

and

$$\mathbf{r}_2 = \mathbf{R} - \frac{m_1}{m_1 + m_2}\mathbf{r}.$$

This is a convenient and useful way of solving for the motion of two particles interacting with a central force. Thus, the problem illustrated in Chapter 3, that of finding the orbit of a particle moving in a central force with fixed center, can be extended to the more realistic problem of two interacting masses.

5.2   The N-Body Problem

Unfortunately, the problem of N mutually interacting particles cannot be reduced to N one-particle problems as is the case for two particles. The fact that the equation of motion of each particle is in general coupled to that of every other particle makes the N-body problem difficult to solve, even numerically, for more than a few particles. However, there are some simplifications that can be made that give important information about the motion of the system, and allow us to solve certain special cases. In particular, suppose we write the position vector of each particle $(\mathbf{r}_i)$ as the sum of the position vector to the center of mass $(\mathbf{R})$ and an "internal" position vector relative to the center of mass $(\mathbf{r}_i')$,

$$\mathbf{r}_i = \mathbf{R} + \mathbf{r}_i'. \tag{5.5}$$

With this choice the momentum, kinetic energy, and angular momentum of a system of particles can each be written as the sum of two parts, one depending on the motion of the center of mass, and the other depending on only the internal part of the motion. See Ref. [15] for details of the derivations.

5.2.1 Momentum

The momentum of a system of particles is almost trivially separated as described above because of the way that the position of the center of mass is defined. That is, since $\mathbf{R}$ is defined by the equation

$$\sum_{i=1}^{N} m_i \mathbf{r}_i = M\mathbf{R} , \tag{5.6}$$

the total momentum of the system is given by

$$\mathbf{P} = \sum_{i=1}^{N} m_i \dot{\mathbf{r}}_i = M\dot{\mathbf{R}}, \tag{5.7}$$

Note that there is no term involving internal coordinates; the total momentum of the system depends on the center of mass velocity, but not the internal velocities.

We proceed further by writing Newton's 2nd law for each of the particles of a system in the form

$$m_i \ddot{\mathbf{r}}_i = \mathbf{F}_i^{int} + \mathbf{F}_i^{ext} , \tag{5.8}$$

where $\mathbf{F}_i^{int}$ and $\mathbf{F}_i^{ext}$ are respectively the forces on the ith particle due to other particles in the system (internal) and objects outside the system (external). By summing the left and right sides of these equations for each particle and applying Newton's 3rd law we obtain an equation that has the form of Newton's 2nd law, but describes a property of the system as a whole:

$$\frac{d\mathbf{P}}{dt} = M\ddot{\mathbf{R}} = \mathbf{F}^{ext} , \tag{5.9}$$

where $\mathbf{F}^{ext}$ is the net external force on the system. Equation 5.9 allows us to justify the previous treatment of extended bodies such as blocks, wedges, *etc.* as particles.

Equation 5.9 can also be used to obtain the law of *conservation of momentum*

> **If the net external force on a closed system of particles is zero, the total momentum of the system remains constant.**

Although the dynamics of the center of mass is described by a simple equation of motion, the center of mass and internal motions are not truly decoupled because, in general, $\mathbf{F}^{ext}$ depends on the internal positions of the particles. In addition, the dynamical equation for $\mathbf{r}'_i$, obtained from Eqs. 5.5 and 5.8,

$$m_i \ddot{\mathbf{r}}'_i = \mathbf{F}^{int}_i + \mathbf{F}^{ext}_i - m_i \ddot{\mathbf{R}} , \qquad (5.10)$$

depends on the motion of the center of mass. However, if external forces are negligible compared to the internal forces, the last two terms of Eq. 5.10 can be neglected, and the internal motion is decoupled from the center of mass motion. Even in this case, however, the internal motion of the particles are coupled to each other since $\mathbf{F}^{int}_i$ depends on the positions of each of the other particles.

5.2.2 Kinetic Energy

The kinetic energy of a system of particles may be written in terms of the center of mass and internal coordinates as

$$K = \sum_{i=1}^{N} \frac{1}{2} m_i (\dot{\mathbf{R}} + \dot{\mathbf{r}}'_i)^2 .$$

Expanding the right side, and using the fact that

$$\sum_{i=1}^{N} m_i \dot{\mathbf{r}}'_i = 0 ,$$

we find that the kinetic energy consists of two parts — one the kinetic energy associated with the motion of the center of mass, and the other describing kinetic energy due to internal motion of the particles:

$$K = \frac{1}{2} \sum_{i=1}^{N} M \dot{\mathbf{R}}^2 + K^{int} , \qquad (5.11)$$

where

$$K^{int} = \frac{1}{2} \sum_{i=1}^{N} m_i \dot{\mathbf{r}}'^2_i . \qquad (5.12)$$

5.2.3 Angular Momentum

The total angular momentum of the system about a point O, which we take to be the origin of the coordinate system, is given by

$$\mathbf{L} = \sum_{i=1}^{N} m_i (\mathbf{r}_i \times \dot{\mathbf{r}}_i) .$$

Replacing $\mathbf{r}_i$ and $\dot{\mathbf{r}}_i$ with their equivalences in terms of center-of-mass and internal coordinates, and expanding, gives four terms inside the sum. Two of them vanish, yielding the separation of angular momentum into center-of-mass and internal parts:

$$\mathbf{L} = M\mathbf{R} \times \dot{\mathbf{R}} + \sum_{i=1}^{N} m_i \mathbf{r}_i' \times \dot{\mathbf{r}}_i' . \tag{5.13}$$

If we take the time derivative of Eq. 5.13, and use Eqs. 5.8 and 5.9 we find

$$\frac{d\mathbf{L}}{dt} = \mathbf{R} \times \mathbf{F}^{ext} + \sum_{i=1}^{N} \mathbf{r}_i' \times \mathbf{F}_i^{ext} + \sum_{i=1}^{N} \mathbf{r}_i' \times \mathbf{F}_i^{int} .$$

The last term in this expression is the sum of torques about the center of mass due to internal forces. It can be shown to be zero if the "strong form" of Newton's 3rd law holds for the internal forces; *i.e.*, if the internal forces between any two particles are not only equal in magnitude and opposite in direction, but also act along the line joining the two particles. An alternative condition for the term to vanish is the assumption that no net work is done by the internal forces in a small virtual rotation about any axis passing through the center of mass. See, e.g., [15] for more discussion of this point.

If the third term is zero, which empirically seems to be the case, we can combine the first two terms to get

$$\frac{d\mathbf{L}}{dt} = \sum_{i=1}^{N} \mathbf{r}_i \times \mathbf{F}_i^{ext} = \tau^{ext} ; \tag{5.14}$$

that is, the time rate of change of angular momentum of a system taken about any point is equal to the net external torque about that point.

From Eq. 5.14 we can derive the law of *conservation of angular momentum* for a system.

> **If the net external torque on a closed system of particles is zero about any point, the angular momentum of the system about that point remains constant.**

5.3 Simple Rigid Body Motion

To make these ideas more concrete, let us look at the special case of a rigid body, a system of particles whose interparticle distances remain fixed. Further, we restrict internal motion to that which can be described as rotation about an axis passing through the center of mass. All particles in a rigid body move in circular paths about the rotation axis, with a common angular speed ω. If we take the rotation axis to be the z-axis for the internal coordinates, the internal kinetic energy simplifies to $K^{int} = \frac{1}{2} I_z \omega^2$; I_z is the moment of inertia about the rotation axis, given by

$$I_z = \sum_{i=1}^{N} m_i(x_i'^2 + y_i'^2) \, . \tag{5.15}$$

Note that I_z is dependent only upon the mass distribution and geometric properties of the rigid body, not upon its dynamics.

The internal part of the angular momentum is similarly simplified when the body is rotating about an axis passing through the center of mass. In particular, since each particle moves in a circle, its velocity is tangential, so that

$$\mathbf{r}_i' \times \dot{\mathbf{r}}_i' = (x_i'^2 + y_i'^2)\omega\hat{\mathbf{z}} \, .$$

The standard right-hand-rule defines which direction along the rotation axis is the positive z-direction. Applying this result to Eq. 5.13 results in the following expression for the angular momentum for a rigid body:

$$\mathbf{L} = M\mathbf{R} \times \dot{\mathbf{R}} + I_z\omega\hat{\mathbf{z}} \, . \tag{5.16}$$

5.3.1   Centers of Mass and Moments of Inertia

The calculation of the center of mass or moment of inertia of a solid body is commonly treated in calculus as an application of multi-dimensional integration. These quantities are important in the dynamics of systems of particles, and particularly for rigid bodies. Although each is defined by sums over the particles that make up the system (Eqs. 5.6 and 5.15), in practice they cannot be calculated by means of these sums because the number of particles is far too large. Thus a continuum approximation is usually made for the mass distribution, and the sums are converted into integrals. With this approximation, each individual mass m_i is replaced by a differential mass element μdV, where μ is the mass density and dV the differential volume element. Thus, in the continuum approximation,

$$\mathbf{R} = \frac{1}{M} \iiint \mathbf{r}\mu \, dV$$

and

$$I_z = \iiint (x^2 + y^2)\mu \, dV \, .$$

In the latter case the moment of inertia has been taken about the z-axis of the coordinate system.

Although the center of mass equation is a vector equation, leading to three separate multiple-integral evaluations, the general approach to calculating $\mathbf{R}$ is the same as that for calculating I_z. We illustrate the process by calculating I_z for an axially-symmetric ellipsoid using Cartesian, cylindrical, and spherical coordinates. A schematic of the solid is shown in Fig. 5.1. For simplicity, we

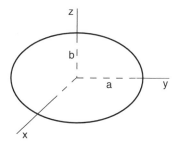

FIGURE 5.1. Ellipsoid with rotational symmetry about z-axis.

take the mass density to be uniform, $\mu = M/V$, where M is the total mass and V is the volume of the ellipsoid. Although this solid is highly symmetric, knowledge of the symmetry is not used to calculate the moment of inertia in order to better illustrate the technique for general, non-symmetric shapes.

Cartesian Coordinates

In Cartesian coordinates, the differential volume element is simply $dV = dx\, dy\, dz$. The difficulty comes in determining the limits on the variables since they are interrelated through the equation for the surface:

$$\frac{x^2 + y^2}{a^2} + \frac{z^2}{b^2} = 1 .$$

We can find the limits to the z-integral in terms of x and y by solving this equation for z. The negative root is appropriate for the lower limit on z, while the positive root is used for the upper limit. Limits on the y-integral can be obtained by solving the equatorial equation,

$$x^2 + y^2 = a^2 ,$$

for y. Again the negative and positive roots apply to the lower and upper limits, respectively. Finally, the limits on the x-integral are $\pm a$. The Maple evaluation is given below.

```
> eq := (x^2 + y^2)/a^2 + z^2/b^2 = 1:
> solve(eq, z);
```

$$\frac{b\sqrt{-x^2 - y^2 + a^2}}{a}, \quad -\frac{b\sqrt{-x^2 - y^2 + a^2}}{a}$$

```
> z1 := "[2]:     z2 := ""[1]:
> solve(x^2+y^2=a^2, y);
```

$$\sqrt{-x^2 + a^2}, \quad -\sqrt{-x^2 + a^2}$$

```
> y1 := "[2]:      y2 := ""[1]:
> V := int( int( int(1, z=z1..z2), y=y1..y2), x=-a..a);
```

$$V := \tfrac{4}{3}\pi b a^2$$

```
> Iz := (M/V)*int( int( int(x^2+y^2, z=z1..z2), y=y1..y2), x=-a..a);
```

$$Iz := \tfrac{2}{5}M a^2$$

Cylindrical Coordinates

In cylindrical coordinates, the volume element is $dV = \rho\, d\rho\, d\phi\, dz$, and $x^2 + y^2 = \rho^2$. The ϕ integral is trivial since the integrand does not depend on ϕ; its range is $0 \le \phi \le 2\pi$. Depending on whether the ρ- or z-integral is evaluated next, we can determine the appropriate limits from the equation for the surface,

$$\frac{\rho^2}{a^2} + \frac{z^2}{b^2} = 1 .$$

The Maple session below solves for z in terms of ρ to get the upper and lower limits on the z-integral in the same way as done in the previous section. The range of the ρ-integral is then $0 \le \rho \le a$. If instead the ρ-integral were done first, the lower limit would be 0, and the upper limit would be the positive root obtained from solving the surface equation for ρ in terms of z. The limits on z would then be $\pm b$.

```
> eq := rho^2/a^2 + z^2/b^2 = 1:
> solve(eq, z);
```

$$\frac{b\sqrt{-\rho^2 + a^2}}{a}, \quad -\frac{b\sqrt{-\rho^2 + a^2}}{a}$$

```
> z1 := "[2]:      z2 := ""[1]:
> V := int( int( int(rho, phi=0..2*Pi), z=z1..z2), rho=0..a);
```

$$V := \frac{4\,(a^2)^{3/2}\,\pi b}{3\quad a}$$

```
> Iz := (M/V)*int( int( int( rho^3, phi=0..2*Pi), z=z1..z2), rho=0..a);
```

$$Iz := \tfrac{2}{5}M a^2$$

Spherical Coordinates

The spherical coordinate volume element is $dV = r^2 dr\, \sin\theta\, d\theta\, d\phi$, and $x^2 + y^2 = r^2 \sin^2\theta$. The equation for the surface,

$$\frac{r^2 \sin^2\theta}{a^2} + \frac{r^2 \cos^2\theta}{b^2} = 1 ,$$

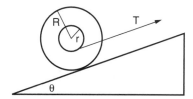

FIGURE 5.2. Schematic of a yo-yo on an incline.

can be used to find the upper limit on r as a function of θ. We must choose the positive root. The θ integral ranges over $0 \le \theta \le \pi$. As with cylindrical coordinates, the ϕ integral is trivial. The Maple sequence is

```
> eq := (r*sin(theta)/a)^2 + (r*cos(theta)/b)^2 = 1:
> solve(eq, r);
```

$$\frac{\sqrt{\sin(\theta)^2 b^2 + \cos(\theta)^2 a^2}}{\left(\dfrac{\sin(\theta)^2}{a^2} + \dfrac{\cos(\theta)^2}{b^2}\right) ab}, \quad -\frac{\sqrt{\sin(\theta)^2 b^2 + \cos(\theta)^2 a^2}}{\left(\dfrac{\sin(\theta)^2}{a^2} + \dfrac{\cos(\theta)^2}{b^2}\right) ab}$$

```
> R := simplify("[1]);
```

$$R := \frac{ab}{\sqrt{b^2 - b^2 \cos(\theta)^2 + \cos(\theta)^2 a^2}}$$

```
> V := int( int( int(r^2*sin(theta), phi=0..2*Pi), r=0..R), theta=0..Pi);
```

$$V := \frac{4}{3} \frac{ba^3 \pi}{\sqrt{a^2}}$$

```
> Iz := (M/V)*int( int( int(r^4*sin(theta)^3, phi=0..2*Pi), r=0..R), theta=0..Pi);
```

$$Iz := \frac{2}{5} Ma^2$$

5.3.2 Yo-Yo on an Incline

As an example of rigid body motion, combining translation of the center of mass with rotation about an axis, let us consider a yo-yo on an inclined plane. The yo-yo has a mass M and is made up of two disks with outer radius of R joined by a short shaft of radius r. The situation is pictured in Fig. 5.2. For simplicity, we assume that the moment of inertia of the yo-yo about an axis passing through the center of the disks and shaft, perpendicular to the page, is dominated by the contribution of the two disks. Thus, we take

$$I_z = \frac{1}{2} MR^2 .$$

We also assume that the tension in the string is maintained in such a way that the yo-yo rolls without slipping on the incline.

Three objects exert forces on the yo-yo: the string, the earth, and the incline. As usual, the force exerted by the incline is separated into normal and frictional components. We choose the fixed (external) coordinate system so that the positive x-axis is parallel to the incline in the upward direction, and the positive y-axis is perpendicular to the incline, also in the upward direction. With these choices, the center of mass motion is described by the component equations

$$T + f - Mg \sin \theta = MA \tag{5.17}$$

and

$$N - Mg \cos \theta = 0 . \tag{5.18}$$

In Eq. 5.17, A is the value of the acceleration of the center of mass, and the sign of f has been chosen as if friction were directed up the plane. Its actual direction depends on the other quantities in the problem; for conditions in which friction is directed down the plane the value of our f is negative.

If we evaluate torques and angular momentum about the center of mass of the yo-yo, the equations for rotational dynamics of a rigid body, Eq. 5.14, together with Eq. 5.16 with $\mathbf{R} = 0$, provides a third equation,

$$Tr + fR = I_z \alpha , \tag{5.19}$$

where $\alpha = \dot{\omega}$.

Finally, the fact that the yo-yo rolls without slipping provides a constraint relating the angular acceleration about the center of mass to the linear acceleration of the center of mass:

$$\alpha = -\frac{A}{R} . \tag{5.20}$$

Note that the minus sign is necessary for consistency in the choice of positive directions for the linear and rotational motions.

These equations can now be solved. We assume that T is provided by a known applied force, and find A, N, and f in terms of T and the other given quantities. The following Maple instructions solve the equations, check the solutions, and assign values to A, N, and f.

```
> eq1 := T + f - M*g*sin(theta) - M*A = 0:
> eq2 := N - M*g*cos(theta) = 0:
> eq3 := T*r + f*R - Iz*alpha = 0:
> alpha := -A/R:
```

> sol := solve({eq1,eq2,eq3}, {N,f,A});

$$sol := \left\{ N = Mg\cos(\theta), \; f = \frac{-Iz\,T + Iz\,Mg\sin(\theta) - Tr\,MR}{Iz + MR^2}, \right.$$

$$\left. A = -\frac{R\,(-TR + Tr + Mg\sin(\theta)\,R)}{Iz + MR^2} \right\}$$

> simplify(subs(sol,{eq1,eq2,eq3}));

$$\{0 = 0\}$$

> Iz := 1/2*M*R^2:
> assign(sol):

N is not affected by T, but f and A clearly are, so we collect terms in T for them:

> collect(A, T, simplify);   collect(f, T, simplify);

The results are

$$A = -\frac{2}{3}g\sin\theta + \frac{2}{3}\frac{(R-r)T}{MR} \tag{5.21}$$

and

$$f = \frac{1}{3}Mg\sin\theta - \frac{1}{3}\frac{(R+2r)T}{R}. \tag{5.22}$$

Consider first the motion when the force applied to the string, represented by T, is zero. In this case the yo-yo accelerates down the plane at a rate $\frac{2}{3}$ of what it would if it slid down without friction. This is because the frictional force directed up the plane is $\frac{1}{3}$ of the component of the gravitational force directed down the plane. (The precise ratio of $\frac{2}{3}$ is due to the specific choice for I_z.) As the applied force is increased, the frictional force decreases, eventually passing through zero and reversing direction. For large T, the yo-yo accelerates up the plane and the frictional force is directed down the plane. Assuming that the coefficient of friction is large enough, this direction for f is necessary in order to provide the torque required for the yo-yo to rotate clockwise and roll up the string. The value for the normal force can be used to obtain the minimum required coefficient of friction.

The kinematics equations for constant acceleration can be used along with Eq. 5.21 to find the center-of-mass velocity of the yo-yo as a function of time or distance rolled. Alternatively, we can find the magnitude of the center-of-mass velocity or the angular speed, as well as the frictional force, by using the linear and rotational forms of the work-energy theorem. Let X be the distance that the yo-yo has travelled up or down the plane, and Φ the total

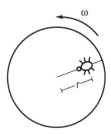

FIGURE 5.3. Schematic of a beetle crawling radially inward on a freely rotating turntable.

angular displacement about the center of mass. The two theorems yield the following equations, assuming that the yo-yo starts from rest:

$$(T + f - Mg \sin \theta)X = \frac{1}{2}MV^2 \qquad (5.23)$$

and

$$(fR + Tr)\Phi = \frac{1}{2}I_z\omega^2 . \qquad (5.24)$$

By adding the condition for rolling without slipping, $\Phi = -X/R$, which also gives $\omega = -V/R$, solve Eqs. 5.23 and 5.24 for f and V. Verify that the result for f is the same as found previously, and show that the solution for V is consistent with that for A in Eq. 5.21.

5.3.3 Beetle on a Turntable

A disk of mass M and radius R rotates freely about an axis perpendicular to its flat surface and passing through its center. Its initial angular speed is ω_0. On the outside edge of the disk is a beetle of mass m that starts crawling toward the center of the disk at a constant speed v_r relative to the disk. Assume that its motion is maintained along a radius of the disk. How will the angular speed of the disk change in time?

The situation is depicted in Fig. 5.3. As the beetle crawls, its distance to the axis of rotation decreases linearly in time, causing the total moment of inertia of the disk-beetle system to decrease. Since the angular momentum of the system is constant, its angular speed must increase to compensate. To solve the problem, we first define the distance of the beetle from the rotation axis and the total moment of inertia as functions of time.

```
> r := t -> R - vr*t:
```

> Iz := t -> 1/2*M*R^2 + m*r(t)^2:

The angular momentum of the system is

> L := Iz(t)*omega(t):

By setting the time derivative of L equal to zero we get the equation of motion for $\omega(t)$:

> eq := diff(L,t) = 0;

$$eq := -2m(R - vr\,t)\,vr\,\omega(t) + \left(\frac{1}{2}MR^2 + m(R - vr\,t)^2\right)\left(\frac{\partial}{\partial t}\omega(t)\right) = 0$$

The solution is obtained and checked with the commands

> sol := dsolve({eq, omega(0)=omega0}, omega(t)):
> simplify(subs(sol,eq));
$$0 = 0$$

> assign(sol):   omega(t);

resulting in

$$\omega(t) = \frac{\omega_0 R^2(M + 2m)}{MR^2 + 2m(R - v_r t)^2} \tag{5.25}$$

Since the beetle travels a finite distance R, t is limited to the range $0 \le t \le R/v_r$.

Obvious limiting cases to verify are $M \to \infty$, $m = 0$, and $v_r = 0$. On physical grounds, what should $\omega(t)$ be in these limits?

To provide additional insight, as well as another check on the solution, we verify the work-energy theorem for the problem. In this context the work done by the radial component of the force on the beetle should equal the change in rotational kinetic energy of the system. The tangential, static frictional force between the beetle and the disk does no net work on the system.

To calculate the work, note that the radial component of the force on the beetle satisfies the equation

$$F_r = m(\ddot{r} - r\omega^2)\,,$$

where $r = R - v_r t$ and ω is given in Eq. 5.25. The work is

$$W = \int_0^t F_r \frac{dr}{dt}\,dt\,.$$

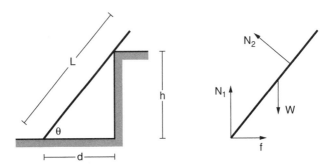

FIGURE 5.4. Schematic of and force diagram for a ladder leaning against a building.

Continuing the dialogue that led to Eq. 5.25, we define functions for $\omega(t)$ and $F_r(t)$ with the commands

```
> Omega := unapply(omega(t), t):
> Fr := t -> m*(diff(r(t),t,t) - r(t)*Omega(t)^2):
```

Now we can find the work and the change in kinetic energy, and verify their equivalence with

```
> Work := int(Fr(s)*diff(r(s),s), s=0..t):
> dK := 1/2*Iz(t)*Omega(t)^2 - 1/2*Iz(0)*Omega(0)^2:
> simplify(Work - dK);
                        0
```

5.4 Equilibrium of a Rigid Body

A rigid body is in equilibrium if there is no net external force on it and the net torque about all points is zero. Under these circumstances the center of mass will experience no linear or rotational acceleration. A static rigid body is obviously a special case of equilibrium.

As an example, we consider a uniform ladder of weight W and length L leaning against a structure whose height is $h < L$. The situation is pictured in Fig. 5.4. The geometry of the system requires $\arcsin(h/L) < \theta < \frac{1}{2}\pi$. There is a static frictional force between the ladder and the ground, but negligible friction between the ladder and the vertical structure. Let us examine the conditions required for equilibrium of the ladder.

The first requirement is that the net external force be zero. The force diagram in Fig. 5.4 leads to two equations. Assuming positive directions upward and to the right, the vertical and horizontal force equations are

$$N_1 + N_2 \cos\theta - W = 0$$

and
$$-N_2 \sin\theta + f = 0 \ .$$

We equate the net torque about the foot of the ladder to zero, with the usual assumption that counterclockwise torques are positive, to get a third equation,
$$-W\left(\frac{L}{2}\right)\sin\left(\frac{\pi}{2}-\theta\right) + N_2\left(\frac{h}{\sin\theta}\right) = 0 \ .$$

These equations can be entered into Maple and solved simultaneously for the three unknown forces f, N_1, and N_2 as we have done before. The results are
$$f = \frac{WL\sin^2\theta\cos\theta}{2h} \ , \tag{5.26}$$

$$N_1 = \frac{W\left(2h - L\sin\theta\cos^2\theta\right)}{2h} \ , \tag{5.27}$$

and
$$N_2 = \frac{WL\sin\theta\cos\theta}{2h} \ . \tag{5.28}$$

The solutions should be checked with Maple and values assigned to the variables with the assign command.

Two simple and worthwhile limiting cases to verify are $W \to 0$ and $\theta \to \frac{\pi}{2}$. In the former, if the ladder has no weight, all of these forces should vanish, as they do. In the latter case, when the ladder is vertical, $N_2 \to 0$ and no friction is required to keep the bottom of the ladder from sliding, so $f \to 0$. The normal force caused by the floor must then cancel the gravitational force on the ladder, resulting in $N_1 \to W$.

What happens if the height of the structure approaches zero? At first glance, this $h \to 0$ limit appears problematical. However, a look back at the figure reveals that as $h \to 0$, $\theta \to 0$ in such a way that $h/\tan\theta$, the distance from the foot of the ladder to the wall, remains fixed. If this distance is d, we get the appropriate $h \to 0$ limits by setting $h = d\tan\theta$ and letting $\theta \to 0$. This is easily done for f with the Maple command

> limit(subs(h=d*tan(theta), f), theta=0);

$$0$$

Similar commands can be applied to get limits for N1 and N2. The $h \to 0$ limits are
$$f = 0 \ ,$$

$$N_1 = W\left(1 - \frac{L}{2d}\right) \ ,$$

and
$$N_2 = \frac{WL}{2d} \ .$$

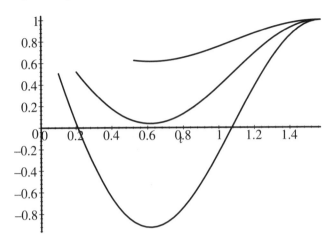

FIGURE 5.5. Plot of the normal force of the floor upon the ladder as a function of θ for several values of the height of the building.

These are the correct expressions for a horizontal ladder of length L supported at two points: at one end and at a distance d from that end. Obviously in this limit $d > \frac{1}{2}L$; otherwise N_1 would have to be negative to enforce equilibrium, which is not possible since the floor cannot pull downward on the ladder.

N_1 must also be positive for the general $\theta > 0$ case. As we see by examining Eq. 5.27, this can lead to restrictions on the range of θ for a given value of h. To see this, we use Maple to plot N_1 as a function of θ for several values of h. First, we define a function of h and θ,

```
> FN1 := unapply(N1/W, h, theta):
```

When given h and θ, this function returns the value of the force N_1 in units of the weight of the ladder. Next, we assign L.

```
> L := 1:
```

This is not a restriction on the problem; rather, it is simply a change of scale that requires h to be interpreted as a fraction of the ladder length. We further define the plotting function

```
> PL := h -> plot(FN1(h,theta), theta=arcsin(h)..Pi/2):
```

to plot N_1 as a function of θ for any allowed value of h $(0 \leq h \leq L)$. To create multiple curves on a single graph we apply the display command, which is in the plots package. This is necessary in this case because the lower limit on each N_1 vs. h plot is different. Figure 5.5 shows the result for $h/L = 0.1$, 0.2, and 0.5. This figure is obtained with the command

> plots[display]({plot(0, theta=0..Pi/2), PL(.1), PL(.2), PL(.5)});

The first plot in the set forces the full range $0 \leq \theta \leq \frac{\pi}{2}$ for the horizontal axis.

As can be seen in the figure, for the two larger values of h, static equilibrium can occur for any value of θ between $\arcsin(h/L)$ and $\frac{\pi}{2}$, provided the coefficient of friction is large enough. (The minimum required coefficient of static friction for given h and θ can be obtained by evaluating f/N_1.) For $h/L = 0.1$, however, there is a range of values for which static equilibrium cannot occur *regardless of how large the coefficient of friction*, since physically N_1 cannot be negative. The approximate range for θ can be read from the figure. To get more accurate values for the end points, we employ Maple's ability to compute numerical values for the roots of an equation.

> fsolve(FN1(.1,theta)=0, theta, 0..0.4);

.2107044771

> fsolve(FN1(.1,theta)=0, theta, 1..1.2);

1.073519922

These θ values are the zero crossings for N_1 when $h/L = 0.1$. The accuracy is controlled by the value of the Maple global variable Digits.

5.5   Coupled Harmonic Oscillators

As a final example of multiple-particle systems we consider two masses, each connected by a spring to a fixed point and connected to each other by a third spring. For simplicity, we take the masses to be equal (m) and the force constants for the springs connected to the walls to be the same (k). The force constant for the interconnecting spring is κ. The setup is shown in Fig. 5.6.

The position coordinates used to describe the motion are the displacements of the two masses from their equilibrium positions. We call these displacements x_1 and x_2. The assumed positive directions are indicated by the arrows in the figure. With these choices, Newton's 2nd law applied to each mass gives the following two equations of motion, assuming no damping forces:

$$-kx_1 - \kappa(x_1 + x_2) = m\ddot{x}_1 , \tag{5.29}$$

$$-kx_2 - \kappa(x_1 + x_2) = m\ddot{x}_2 . \tag{5.30}$$

Dividing each side by m and collecting terms yields

$$\ddot{x}_1 + \omega_0^2 x_1 + \omega_1^2(x_1 + x_2) = 0 , \tag{5.31}$$

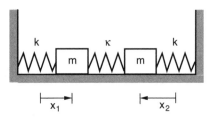

FIGURE 5.6. Two masses connected by springs to walls and each other.

$$\ddot{x}_2 + \omega_0^2 x_1 + \omega_1^2(x_1 + x_2) = 0 \,, \tag{5.32}$$

where the quantities ω_0 and ω_1 are given by

$$\omega_0 = \sqrt{\frac{k}{m}}$$

and

$$\omega_1 = \sqrt{\frac{\kappa}{m}} \,.$$

We enter Eqs. 5.31 and 5.32 into a Maple session, labelling them **eq1** and **eq2**, and request Maple to solve them subject to initial conditions $x_1(0) = x_{10}$, $\dot{x}_1(0) = v_{10}$, $x_2(0) = x_{20}$, and $\dot{x}_2(0) = v_{20}$. The task is done with the commands

```
> assume(omega0>0);   assume(omega1>0);
> sol := dsolve({eq1, eq2, x1(0)=x10, D(x1)(0)=v10,
>          x2(0)=x20, D(x2)(0)=v20}, {x1(t), x2(t)});
```

With minor manual simplification, the following are the solutions returned by Maple:

$$x_1(t) = \tfrac{1}{2}(x_{10} - x_{20})\cos(\omega_- t) + \tfrac{1}{2}\frac{v_{10} - v_{20}}{\omega_-}\sin(\omega_- t)$$
$$+ \tfrac{1}{2}(x_{10} + x_{20})\cos(\omega_+ t) + \tfrac{1}{2}\frac{v_{10} + v_{20}}{\omega_+}\sin(\omega_+ t) \,, \tag{5.33}$$

$$x_2(t) = -\tfrac{1}{2}(x_{10} - x_{20})\cos(\omega_- t) - \tfrac{1}{2}\frac{v_{10} - v_{20}}{\omega_-}\sin(\omega_- t)$$
$$+ \tfrac{1}{2}(x_{10} + x_{20})\cos(\omega_+ t) + \tfrac{1}{2}\frac{v_{10} + v_{20}}{\omega_+}\sin(\omega_+ t) \,, \tag{5.34}$$

with

$$\omega_- = \sqrt{\frac{k}{m}} \tag{5.35}$$

and

$$\omega_+ = \sqrt{\frac{k + 2\kappa}{m}} \, . \tag{5.36}$$

The only difference between the two solutions is that $x_1(t)$ and $x_2(t)$ have opposite signs for the $\cos(\omega_- t)$ and $\sin(\omega_- t)$ terms.

This problem provides a good example of how a little planning can help you encourage Maple to provide more convenient output. Rearranging Eqs. 5.29 and 5.30, replacing k/m and κ/m with ω_0^2 and ω_1^2 to get Eqs. 5.31 and 5.32, and declaring ω_0 and ω_1 positive coaxes Maple to return solutions in terms of sin and cos functions as indicated. If, however, you ask Maple to solve Eqs. 5.29 and 5.30 directly, with the same dsolve call, the solutions are returned as exponential functions with imaginary arguments. To get a form similar to that shown for $x_1(t)$ and $x_2(t)$, convert the exponentials to trigonometric functions, and then collect similar terms together.

A noteworthy characteristic of the solutions for $x_1(t)$ and $x_2(t)$ is that each consists of a sum of sinusoidal terms of two distinct frequencies. Thus we expect a plot of $x_1(t)$ or $x_2(t)$ vs. t will show a "beat" phenomenon for the amplitudes of the position. To illustrate this effect, we assign values to the parameters, and define functions for the positions and energies of the oscillators. Note that since the variables omega0 and omega1 were assumed positive earlier, we must use an assign command to give them particular values.

```
> assign(sol):
> k := m*omega0^2:    kappa := m*omega1^2:
> v10 := 0:   v20 := 0:   m := 1:
> assign(omega0 = 1):
> assign(omega1 = 1/2):
> X1 := unapply(x1(t), x10, x20, t):
> X2 := unapply(x2(t), x10, x20, t):
> V1 := unapply(diff(x1(t),t), x10, x20, t):
> V2 := unapply(diff(x2(t),t), x10, x20, t):
> U := (x10, x20, t) -> 1/2*k*X1(x10,x20,t)^2 + 1/2*k*X2(x10,x20,t)^2
>           + 1/2*kappa*( X1(x10,x20,t) + X2(x10,x20,t) )^2:
> K := (x10, x20, t) -> 1/2*m*V1(x10, x20, t)^2
>           + 1/2*m*V2(x10, x20, t)^2:
> En := (x10, x20, t) -> K(x10, x20, t) + U(x10, x20, t):
```

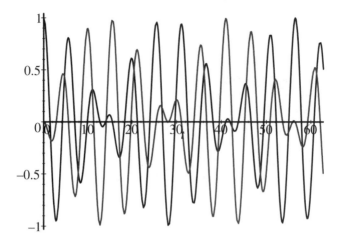

FIGURE 5.7. Positions of the two coupled masses as a function of time.

Figure 5.7 is the result of plotting the displacements of the two masses when started from rest, subject to the conditions $x_1(0) = 1$ and $x_2(0) = 0$. This figure is obtained with the Maple command

```
> plot({X1(1,0,t), X2(1,0,t)}, t=0..20*Pi);
```

It clearly shows the "beat" effect as the peak displacement of each mass increases and decreases. Moreover, we see that the beats are out of phase, as one mass is oscillating with maximum amplitude when the other is scarcely moving. Energy is clearly being swapped back and forth between the two masses, although because of the potential energy in the coupling spring, the total energy cannot be clearly divided between the two masses.

It is worthwhile to look at the variation of the energy of the system for the same initial conditions. Figure 5.8 shows the kinetic, potential, and total energies of the system of masses and springs as functions of time. The figure is obtained with the command

```
> plot({K(1,0,t), U(1,0,t), En(1,0,t)}, t=0..10*Pi);
```

One thing to note from this figure is that the kinetic and potential energies are $180°$ out of phase; i.e., when one is at a relative maximum the other is at a relative minimum. Like the displacements shown in Fig. 5.7, the kinetic and potential energies also undergo beats. Interestingly, their beats are in phase; that is, the envelopes of the kinetic energy and the potential energy curves increase and decrease together about a mean positive value of about 0.33 in the units chosen. As expected, however, the total energy is constant throughout the motion.

Looking back at the solutions in Eqs. 5.33 and 5.34, we notice that by an

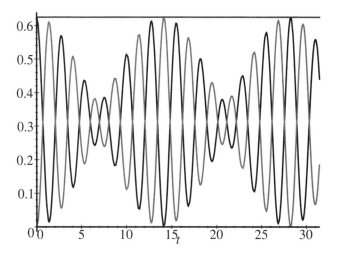

FIGURE 5.8. Kinetic, potential, and total energies as a functions of time for the coupled harmonic oscillators.

appropriate choice of initial conditions each of the two masses can be made to oscillate together at one or the other of two frequencies, ω_+ or ω_-. For example, if $x_{10} = x_{20}$ and $v_{10} = v_{20}$, then $x_1(t) = x_2(t)$ and the two masses move toward or away from each other in synchrony, at an angular frequency of ω_+. On the other hand, if $x_{10} = -x_{20}$ and $v_{10} = -v_{20}$, the two masses oscillate synchronously in the same direction at an angular frequency of ω_-.

We can use animate to see these two special collective modes of oscillation, as well as more general cases. For simplicity, we take the initial velocity of each mass to be 0 and the initial position of the first mass to be 1.

```
> v10 := 0:   v20 := 0:   x10 := 1:
```

Let us assume the equilibrium positions of the two masses are -1 and 1. The Maple commands below display the opposite-direction motion; Tplus is the period of the motion.

```
> x20 := x10:   Tplus := 2*Pi/sqrt(omega0^2+2*omega1^2):
> plots[animate]({x1(t)-1, x2(t)+1}, s=0..1, t=0..Tplus, frames=20);
```

The s=0..1 is necessary because animate requires two variable ranges in its argument list. Since x1(t) and x2(t) do not depend on the s variable, the position of the two masses are represented by horizontal lines on the graph. For best observation the frames should be interactively set to cycle. Also, the display rate of the animation may need to be adjusted to clarify the motion. With this setup one horizontal line (representing m_1) oscillates vertically about the -1 position, and a second line (representing m_2) oscillates about the +1 position.

The same-direction motion can be observed with a similar animate command after changing the value of x20 and defining the period of the motion.

```
> x20 := -x10:      Tminus := 2*Pi/omega0:
> plots[animate]({x1(t)-1, x2(t)+1}, s=0..1, t=0..Tminus, frames=20);
```

Different motions can be shown by calling animate after setting x20 to other values in the range [-1,1].

The two special sets of solutions, where the masses oscillate synchronously at one of the angular frequencies given in Eqs. 5.35 and 5.36 are known as *normal modes*. They can be described by the following two linear combinations of $x_1(t)$ and $x_2(t)$:

$$
\begin{aligned}
x_+(t) &= x_1(t) + x_2(t) \\
&= x_+(0)\cos(\omega_+ t) + \frac{\dot{x}_+(0)}{\omega_+}\sin(\omega_+ t)
\end{aligned}
$$

and

$$
\begin{aligned}
x_-(t) &= x_1(t) - x_2(t) \\
&= x_-(0)\cos(\omega_- t) + \frac{\dot{x}_-(0)}{\omega_-}\sin(\omega_- t) .
\end{aligned}
$$

For more complex coupled oscillations the normal modes are more difficult to find. There are established techniques for doing so which can be found in advanced classical mechanics texts. The importance of normal modes lies in the fact that they describe simple collective motions that can be used in linear combinations to construct mathematically any allowed motion of the system. See references for details.

5.6   Problems

1. A solid cylinder of mass M and radius R rolls without slipping down a fixed incline. The angle of the incline is θ, and the cylinder starts from rest.

   (a) Find the speed of the center of mass of the cylinder after it has rolled a distance L down the incline using

      i. conservation of energy,
      ii. Newton's 2nd law and the torque-angular acceleration relation.

   (b) Verify the $R \to 0$ and $\theta \to 0$ limiting cases. Why doesn't the $\theta \to \frac{\pi}{2}$ limit go over to a free fall result?

   (c) For given θ, find the minimum value for the coefficient of static friction required to maintain rolling without slipping.

2. A yo-yo consists of two disks of radius R and total mass M joined by a short cylindrical shaft through the disk centers. The shaft has radius r and negligible mass. It is supported by a string which is wound around the shaft. The yo-yo is released from rest and begins to turn as the string unwinds. You may assume that the string is vertical as the yo-yo falls.

   (a) Using Newton's 2nd law and the torque-angular acceleration equation, find the tension in the string and the acceleration of the center of mass of the yo-yo.

   (b) From this acceleration find the speed of the center of mass after the yo-yo has fallen a distance d.

   (c) Use conservation of energy to find the speed of the center of mass after the yo-yo has fallen a distance d. Make sure this calculation agrees with the previous result.

3. Four bricks, each of weight W and length L, are placed on top of one another so that part of each brick extends beyond the one underneath.

   (a) Write down the equations for equilibrium that must be satisfied if each of the top three bricks extends a maximum amount beyond the one below. (Hint: Just at the critical point the force that a brick exerts on the one above acts at the upper edge that is in contact with the supported brick.)

   (b) Solve the equations simultaneously for the amount of maximum extension and the magnitude of the forces between bricks.

(c) Obtain and justify an expression for the maximum amount that the nth brick in a stack can extend beyond the $(n + 1)$st brick (counting from the top). Although you do not need to do so, the correct result can be proved by mathematical induction.

4. A uniform ladder of length L and mass M leans against the vertical wall of a house, making an angle θ with the horizontal. The coefficient of friction between the ground and the bottom of the ladder is μ_1, and that between the wall and the top of the ladder is μ_2.

   (a) Find the static frictional force between the ladder and the ground as a function of θ.

   (b) Verify the $\theta \to \frac{\pi}{2}$ limit of the above result.

   (c) Find the smallest allowed value for θ such that the ladder does not slip.

5. The equation of motion for a rocket moving against a uniform gravitational field g is given by

$$m\frac{dv}{dt} = -u\frac{dm}{dt} - mg \, ,$$

where u is the speed of the expended gases relative to the rocket, and dm/dt is the rate at which expended fuel is ejected.

   (a) Assuming that the mass is ejected at a constant rate,

$$\frac{dm}{dt} = -r \, ,$$

   find the speed of the rocket as a function of time for initial conditions $v(0) = 0$ and $m(0) = M$.

   (b) Expand the expression for the speed for small times and obtain the condition on u and r required to overcome the gravitational field strength.

   (c) Assign values to the parameters and plot $v(t)$. Explain the sudden rise in the curve as $t \to M/r$.

6. A star of mass m is initially moving freely with a speed of v_0 with respect to an inertial reference frame. If not for gravity it would continue moving in a straight line, passing within a distance b of a second star of mass M. The second star is initially at rest with respect to the same reference system. The gravitational force, of course, causes both stars to change their state of motion.

(a) Transform the problem into an effective one-body problem and find the asymptotic angle through which the reduced-mass object will be deflected when its motion is again nearly a straight line.

(b) Transform back to the original coordinates and find the final velocities of both stars, assuming that the initial velocity v_0 is along the x-axis.

References

[1] Milton Abramowitz and Irene A. Stegun, eds., *Handbook of Mathematical Functions*. New York: Dover Publications, Inc.

[2] Ralph Baierlein, *Newtonian Dynamics*. New York: McGraw-Hill, Inc. (1983)

[3] Nancy R. Blachman and Michael J. Mossinghoff. *Maple V Quick Reference*. Pacific Grove, CA: Brooks/Cole Publishing Company (1994)

[4] Bruce W. Char, Keith O. Geddes, Gaston H. Gonnet, Benton L. Leong, Michael B. Monagan, and Stephen M. Watt. *First Leaves: A Tutorial Introduction to Maple V*. New York: Springer-Verlag (1992)

[5] Bruce W. Char, Keith O. Geddes, Gaston H. Gonnet, Benton L. Leong, Michael B. Monagan, and Stephen M. Watt. *Maple V Language Reference Manual*. New York: Springer-Verlag (1991)

[6] Robert M. Corless. *Essential Maple*. New York: Springer-Verlag (1995)

[7] Wade Ellis, Jr., Eugene W. Johnson, Ed Lodi, Daniel Schwalbe. *Maple V Flight Manual*. Pacific Grove, CA: Brooks/Cole Publishing Company (1992)

[8] A. P. French. *Newtonian Mechanics*. New York: W. W. Norton & Company, Inc. (1971)

[9] Ronald L. Greene. "On Integrating Computers into the Physics Curriculum," in *Mathematical Computation with Maple V: Ideas and Applications*, ed. by Thomas Lee. Boston: Birkhäuser (1993)

[10] André Heck. *Introduction to Maple*. New York: Springer-Verlag (1993)

[11] David Hestenes. *New Foundations for Classical Mechanics*. Dordrecht: D. Reidel Publishing Company (1986)

[12] Daniel Kleppner and Robert J. Kolenkow. *An Introduction to Mechanics*. New York: McGraw-Hill, Inc. (1973)

[13] Jerry B. Marion and Stephen T. Thornton. *Classical Dynamics of Particles & Systems, 3rd ed.* San Diego: Harcourt Brace Jovanovich, Publishers (1988)

[14] Darren Redfern. *The Maple Handbook, Maple V Release 3.* New York: Springer-Verlag (1994)

[15] Keith R. Symon. *Mechanics, 3rd ed.* Reading, MA: Addison-Wesley Publishing Company (1971)

Index